T0201941

THE DILEMMAS OF
WONDERLAND

THE DILEMMAS
OF *WONDERLAND*

Decisions in the Age of Innovation

Yakov Ben-Haim

Technion—Israel Institute of Technology

OXFORD
UNIVERSITY PRESS

OXFORD
UNIVERSITY PRESS

Great Clarendon Street, Oxford, OX2 6DP,
United Kingdom

Oxford University Press is a department of the University of Oxford.
It furthers the University's objective of excellence in research, scholarship,
and education by publishing worldwide. Oxford is a registered trade mark of
Oxford University Press in the UK and in certain other countries

First Edition published in 2018

Impression: 1

Published in the United States of America by Oxford University Press
198 Madison Avenue, New York, NY 10016, United States of America

British Library Cataloguing in Publication Data

Data available

Library of Congress Control Number: 2018937425

ISBN 978–0–19–882223–3
DOI: 10.1093/oso/9780198822233.001.0001

Printed and bound by
CPI Group (UK) Ltd, Croydon, CR0 4YY

For Miriam

לאשתי מרים

Contents

Dreams of Wonderland from long ago
Still are strong and near.
Though future far we cannot know,
Our hope fights doubt and fear.

Hopes and dreams are not enough.
Action! Innovation!
But urgent choice is often rough
On calm deliberation.

Good is good, so better seems best,
But surprise and wonders lurk.
Innovations have promise but none the less,
Being new, they may not work.

Dreams of Wonderland from long ago,
The heart should not reject.
But the mind must lead us, as we go,
And against surprise protect.

1

Introduction

Alice ran after the White Rabbit and tumbled into his rabbit
hole, falling gently down, down, until she found herself
in Wonderland. Alice was too big to go through the small
door leading "into the loveliest garden you ever saw" so
she wanted to shrink "like a telescope.... For, you see, so
many out-of-the-way things had happened lately, that Alice
had begun to think that very few things indeed were really
impossible." She found a bottle labelled "DRINK ME" that
might just do it. It wasn't labelled "poison" so Alice tried
it and it worked, though she worried that the shrinking
"might end, you know,... in my going out altogether, like
a candle."[1]

Like Alice, we live in a world of "endless discovery", so "Dive right
in and follow what you love", whether it's smartphones, computers, Internet, or indeed "everything tech related" as declared in a
recent Google Plus ad.

Innovations provide new and purportedly better opportunities, but—because of their newness—they are often more
uncertain and potentially worse than existing options. There are
new drugs, new energy sources, new foods, new manufacturing

---

[1] Material in quotation marks is from Lewis Carroll, (Charles Lutwidge Dodgson), 1865, *Alice's Adventures in Wonderland,* chapter 1, Project Gutenberg, Salt Lake
City, UT, last updated July 14, 2014.

The Dilemmas of Wonderland: Decisions in the Age of Innovation. Yakov Ben-Haim.
© Yakov Ben-Haim 2018. Published in 2018 by Oxford University Press.
DOI: 10.1093/oso/9780198822233.001.0001

technologies, new toys and new pedagogical methods, new weapon systems, new home appliances, and many many other discoveries and inventions.

To use or not to use a new and promising but unfamiliar and hence uncertain innovation? That is the dilemma facing policy-makers, engineers, social planners, entrepreneurs, physicians, parents, teachers; in fact just about everybody in their daily lives. Furthermore, the paradigm of the innovation dilemma characterizes many situations even when a new technology is not actually involved. The dilemma arises from new attitudes, like individual responsibility for the global environment, or new social conceptions, like global allegiance and self-identity transcending all nation-states. Even the enthusiastic belief in innovation itself as the source of all that is good and worthy entails a dilemma of innovation. These dilemmas have far-reaching implications for individuals, organizations, and society at large.

An innovation's newness and its promise for improvement is the source of the dilemma. Tomorrow we will understand the innovation better—its dangers as well as its benefits—but today we must decide. Without our optimistic belief in the future we would remain always in the past. But optimism without under-standing can be dangerous because the innovation may harbour unanticipated and unpleasant surprises. We need a sensible and responsible method for responding to the endless flow of discov-ery and invention.

This book presents a method for managing the challenge of innovation dilemmas and for thinking about their implications. Central to the discussion is the idea of an information gap: the dis-parity between what you *do know* and what you *need to know* in order to make a responsible decision. Our treatment of innovation di-lemmas is based on info-gap theory. However, before we can fully understand the method and its implications, we must first exam-ine two concepts: the boundlessness of our ignorance and the limitation of our ability to achieve optimal outcomes. These two ideas—uncertainty and the limits of outcome-optimization—are linked because outcome-optimization becomes less feasible as uncertainty grows. Much of this book is a study of the interdependence between uncertainty (meaning lack of knowledge) and optimization of outcomes.

After examining a wide range of innovation dilemmas in Chapter 2, we proceed in Chapter 3 to discuss uncertainty, ignorance, and surprise. The converse of the endless possibility for discovery and invention is the boundlessness of our current ignorance about what does or might exist, or about how things work, or how they are changing or will change in the future. We will explore situations in which our understanding of the underlying processes is fragmentary or wrong, and situations involving change or innovation that we cannot anticipate. In these situations, our ability to predict the outcome of our decisions is significantly limited.

Many of the ideas developed in this book rest firmly on the work of two towering scholars of uncertainty: Frank Knight and Herbert Simon.

Knight introduced the distinction between probabilistic risk and what he called "true uncertainty". This latter category, which has come to be known as non-probabilistic Knightian uncertainty, derives from the human potential for discovery and innovation, and underlies our focus on boundless info-gaps. Knightian uncertainty is important and prevalent because of an inherent indeterminism in human affairs that is discussed, separately, in the work of Karl Popper and G. L. S. Shackle.

Simon introduced the concept of bounded rationality: the limited practical ability of individuals and organizations to process all the information that is available to them and to reach optimal solutions to the problems they face. Instead of optimizing, Simon suggested that agents satisfice: achieve good enough solutions to the challenges at hand. Discussion of Simon's concept of satisficing is deferred to Chapter 5.

We conclude our discussion of uncertainty by exploring specific issues raised by boundlessness of the unknown, including linguistic uncertainty, the impossibility of a unified theory of uncertainty, and the possibility that the search for scientific truth will end.

The second element, in addition to uncertainty, that often exacerbates an innovation dilemma is the deep-rooted tendency to persistently seek optimal outcomes. This propensity, and its limitations, are discussed in Chapter 4. The wonders of human progress owe their origin to our thirst for improvement. However, our ability to predict the outcome quality of the options among which we must choose is limited when our knowledge and understanding are severely curtailed. We simply can't know which option will be optimal. Furthermore, seeking "the best" often becomes a moral imperative or an idealistic or rhetorical device, regardless of substantive needs and values. In these cases, optimizing the outcome becomes the goal, rather than making a responsible decision.

After discussing boundless info-gaps and the limits of optimization in Chapters 3 and 4, we are ready to discuss info-gap methods for managing innovation dilemmas, in Chapter 5. The main info-gap approach to managing an innovation dilemma is to make a decision that robustly satisfies critical requirements. The idea is to identify outcomes or consequences that are essential (that must be achieved) and then to choose the alternative that will achieve these outcomes as reliably as possible despite our deep uncertainty about the situation and how our decision will work out.

Focusing on reliable achievement of critical goals is different from focusing on achieving the best possible outcome. The considerations differ and the chosen decision may (or may not) differ. Attempting to achieve the best possible outcome can result in choosing an option that is putatively optimal but nonetheless highly vulnerable to error in our understanding. This can potentially result in a highly unsatisfactory outcome. By focusing on the optimal outcome we ignore the central importance of reliably achieving critical results in the face of severe uncertainty. In contrast, focusing on robustness against our ignorance, while aiming at specific critical goals, enables the decision maker to balance the quality of the outcome with the confidence to achieve an outcome that is acceptable. It is a major procedural error to try to optimize the outcome of the decision when our ignorance severely limits our power to predict it. We conclude Chapter 5 by discussing several examples of the robustness analysis of decisions made under severe uncertainty.

Chapter 6 explores cultures of innovation and progress. Progress sometimes results from path-changing innovation and sometimes from gradual incremental change. These two mechanisms themselves can constitute an innovation dilemma. Revolutionary innovation is sometimes more promising than evolutionary incrementalism. However, revolutions can be vastly uncertain and can develop in surprising and undesirable ways. It may therefore seem more promising to follow a gradual evolutionary path to progress. However, tiny changes are hard to follow, so evolution can meander in unintended and unwanted directions.

Progress—both personal and social—depends on cultural attitudes toward innovation and initiative. Individualism is the preeminent attitude supporting innovation and improvement. Many modern societies devote extensive resources to educating vast cohorts of individuals for initiative and creativity. However, an innovation dilemma is inherent in a system of mass education for nurturing an exceptional elite of innovative individuals. Furthermore, the hallmark and pinnacle of scholarship is the search for universal principles and truths, but this entails an innovation dilemma when the domain of future application of a new proposition is uncertain. These paradoxes create fragility in innovative societies in two ways. Innovation may undermine the shared values that facilitate innovation, and innovation may undermine the institutional basis for innovation. The drive to innovate brings with it many unavoidable dilemmas.

We all face innovation dilemmas, some seemingly modest, like choosing (or refusing) a new computerized device, some more grandiose, like selecting a strategy to reduce global warming. Not all change is progress, and often we won't know whether it is or not until we've tried it. But adopting radically new ideas or opportunities, perhaps a religious revelation or space travel, may change us in ways we cannot anticipate and may even regret (or would have regretted before we changed). The uncertainties are profound and persistent; the more we know, the more we are able to learn, discover, and invent. The explosion of knowledge blasts open even greater vistas of the unknown. The dilemmas of *Wonderland* are all around us. Like Alice, we needn't resist them, but we must proceed wisely.

2

Innovation Dilemmas: Examples

"Alice noticed with some surprise that the pebbles were all turning into little cakes as they lay on the floor, and a bright idea came into her head. 'If I eat one of these cakes,' she thought, 'it's sure to make *some* change in my size; and as it can't possibly make me larger, it must make me smaller, I suppose.'"[1]

Pebbles to little cakes, mute silicon to intelligent robots, even swords to plowshares continue to amaze us (or may someday). Innovation dilemmas come in myriad forms, arising from the human potential for discovery and invention. These innovations are attractive, and promise to be improvements. However, they are distilled from the endless unknown and are accompanied by considerable uncertainty, so their promise may be illusory. This chapter examines a range of innovation dilemmas, selected for their diversity; some are concrete and technological, others abstract and conceptual.

E-Reading

iPads and other brands of tablets are versatile interactive graphic devices. They display pictures in vivid color, project voices with

[1] Lewis Carroll, *Alice's Adventures in Wonderland,* chapter 4.

The Dilemmas of Wonderland: Decisions in the Age of Innovation. Yakov Ben-Haim.
© Yakov Ben-Haim 2018. Published in 2018 by Oxford University Press.
DOI: 10.1093/oso/9780198822233.001.0001

great fidelity, show movies, link to user-chosen options, store personal data, communicate with other devices and link the user to the world. "iPad is transforming the way teachers teach and students learn", declares the Apple website.[2]

Is the iPad a good device for teaching young children reading and other language skills? The possibilities are great: they are portals to vast libraries and learning apps, and they are fun to use. Kids are captivated by the graphic capabilities for storytelling: dogs bark, queens dance and the screen is always active. "But does this count as story time? Or is it just screen time for babies?" asks Douglas Quenqua of the *New York Times*.[3] People have been learning to read for ages, and educational psychologists have deep understanding of the processes of that learning. But reading from an iPad is different from reading from a printed book. That's precisely the attractive promise of this new technology. But are all the differences beneficial? Does the iPad encourage dialog with parents or teachers, or does it isolate children in virtual realities of their own? Do the hyperlinks and graphic virtuosity distract the child from understanding what is being read, or do they enhance engagement with the material? Does the iPad reduce the quality time parents spend with their children, or does it facilitate more diverse parent–child interaction by releasing parents from continual supervision of the child?

"Part of the problem is the newness of the devices. Tablets and e-readers have not been in widespread use long enough for the sorts of extended studies that will reveal their effects on learning."[4] To use or not to use this new device is an innovation dilemma. The new digital device has tremendous purported and evident advantages over the tried-and-true printed book. Its difference is its attraction, but also the source of concern because the impacts of those differences may be far worse than anticipated. The long-term impacts are poorly understood or not known at all because the device is new. At some point in the future the dilemma will vanish because we'll know as much about e-reading as we now know about print-reading. But today we face an innovation dilemma.

Military Hardware: Messerschmitt 262

Aircraft propulsion systems come in many designs, but two distinct concepts are the rotating propeller and jet propulsion. The propeller system uses combustion cylinders (like in a car) that drive pistons whose motion is translated into the rotation of a propeller. The propeller pushes air backwards, and this drives the airplane forward. The jet engine, in contrast, has no propeller; the airplane is driven forward by the backward ejection of gas at high speed, like an air-filled party balloon that is released before the mouth piece is tied. Propeller planes pre-dated jet planes.

German engineers recognized the potential advantages of the greater speed of the jet plane several years before World War II. However, the development of the Messerschmitt Me 262 was delayed and it was introduced operationally into the Luftwaffe only in April 1944.[5] The course of the air war could have run quite differently had the Nazis developed the jet plane earlier.[6] The decision facing the German high command—to stay with conventional propeller planes or to develop and produce the jet plane—is a paradigmatic innovation dilemma.

[5] Walter J. Boyne, 2008, "Goering's Big Bungle", *Air Force Magazine,* 91(11), November 2008, pp. 58–61.

[6] Yehoshafat Harkabi, 1990, *War and Strategy,* (in Hebrew), Maarachot Publishers, Israel Ministry of Defense, p. 470.

The potential advantage of the jet plane was obvious: its greater speed promised to make it invulnerable to interception by conventional propeller planes. This would allow Nazi air superiority both in defensive interception of allied aircraft and in offensive bombing of allied targets. The jet plane was purportedly the better option, rather than the conventional propeller plane. However, technical challenges remained open that might delay production, especially the search for an alloy whose melting point was high enough to resist the high temperatures of the jet engine. Thus the jet plane was more uncertain than the propeller plane, and could possibly lead to a worse outcome: wasted time and resources in the failed development of the jet engine. Hence, the dilemma.

The root of the dilemma is that the innovation—jet propulsion—is purportedly better but more uncertain and thus potentially worse than the alternative (propeller propulsion). This sort of dilemma commonly results from technological innovation. Innovations, inventions, and discoveries in many disciplines result in the same dilemma.

Bipolar Disorder and Pregnancy

The psychiatrist Dr Vivien Burt and her colleagues wrote that "Bipolar disorder poses uniquely gender-specific challenges for women considering the health and well-being of their unborn children." The challenges arise from "unpredictable but not unexpected complexities". A stark innovation dilemma faces the bipolar woman who wants to become pregnant with the "safest medications possible" for her and her unborn child.[7]

Type I bipolar disorder is a clinical psychiatric condition characterized by the occurrence of one or more manic episodes or

[7] Vivien Burt, Caryn Bernstein, Wendy S. Rosenstein, and Lori L. Altshuler, 2010, "Maintaining psychiatric stability in the real world of obstetric and psychiatric complications", *American Journal of Psychiatry,* 167:8, pp. 892–897, pp. 892–3.

mixed manic-depressive episodes. Three concepts are needed to appreciate type I bipolar disorder: mania, depression, and their mixture.

A manic episode is an extended period (typically seven days or more) of unusually and continuously effusive and open elated or irritable mood, where the mood is not induced by drugs or medications or other illness. Mania is an intense feeling of euphoria and self confidence, even of personal indestructibility, and is at the opposite extreme from depression. Mania can lead to rash, irresponsible, and dangerous behavior.

A major depressive episode entails loss of interest in or pleasure from daily activities, as well as feelings such as sadness, anxiety, emptiness, hopelessness, helplessness, worthlessness, guilt or irritability, loss of appetite or overeating, problems concentrating, problems remembering details or making decisions, and thoughts of or attempted suicide.

A mixed state combines mania and depression, with simultaneous occurrence of agitation, anxiety, fatigue, guilt, impulsiveness, irritability, morbid or suicidal thoughts, panic, paranoia, and rage.

A woman with type I bipolar disorder faces a complex combination of medical and psychiatric challenges in conceiving and bearing a healthy child while also maintaining her own psychiatric well-being. Clinical guidelines assist the woman and her clinicians in managing the pregnancy. However, standardized guidelines are often not suited to a specific woman's constellation of psychiatric and medical issues. These may include medical conditions that exist simultaneously with and independently of the bipolar disorder. These comorbid conditions can include hypothyroidism and polycystic ovary syndrome, or obstetric complications such as gestational diabetes.

Together with the basic condition of bipolarism, this constellation of issues requires the use of a collection of medications. Some essential psychiatric drugs are, or may be, teratogenic, so their use must be in limited dosage (if used at all). Furthermore, the interactions between different drugs are highly uncertain because of patient-specific response. Uncertainty also arises because "clinicians and patients are often faced with newly published data that is [sic] inconsistent with prior data." In addition, pharmacological stabilization of bipolar disorder is a slow process and may take years to achieve. Burt *et al.* conclude that "Lack of data, unpredictable response to multiple medications, and potential complications required that the treatment team be flexible in their plan, tolerate uncertainty, and educate the patient and her husband about known and potential risks at every step along the way."[8]

A bipolar woman, who has reached psychiatric stability after an extended period of pharmacological balancing, faces a stark innovation dilemma. On one hand she can refrain from pregnancy, continue her stabilizing medications, and forego both the joys of parenthood and the risks to herself and her child of complications or injury during pregnancy. This might be viewed as the standard or conventional attitude; "common sense" based on long experience with standard therapies and medications. On the other hand, new medications and therapeutic approaches have become available that offer the possibility of bearing a healthy child without unduly prolonged adverse psychiatric impact on the mother. Indeed, new drugs and new data will probably become available even during the course of the pregnancy. These promising pharmacological and therapeutic innovations have substantially uncertain outcomes precisely because of their newness and because of their unknown impact on the specific woman's constellation of medical and psychiatric issues.

In discussing a specific clinical case, Dr Burt and her colleagues wrote that the patient, "Ms. M", "wanted to become pregnant

[8] See Burt *et al.*, 2010, pp. 892, 896.

with the safest medications possible" despite psychiatric and physiological risks to the mother and possible birth defects in the child. After much deliberation, a complex combination of medications was used to balance diverse psychiatric and physiological factors and "to maximize mood stability."[9]

Dr Burt and her colleagues report that Ms. M maintained psychiatric health through continuous professional care and intervention, and bore a healthy child who, "At 29 months, . . . , is developing steadily and is a happy, articulate, emotionally stable, responsive, and cognitively bright toddler."[10] However, this happy outcome was unknowable when Ms. M had to make her choice between her status quo and the possibility of benefiting from innovative but unfamiliar and uncertain pharmacological developments.

Disruptive Technology: The Manager's Innovation Dilemma

Many firms have successfully maintained their financial standing by carefully monitoring and responding to their customers' needs, and by actively investing in new technologies or methods that sustain and extend the utility of their products to those customers. For example, Christensen[11] discusses the continued improvement in the data density of computer disk drives. Ferrite read/write heads provided continual improvement from 1975 to 1990, being overtaken by thin-film heads, and then by magneto resistive heads, that continued the improvement of computer performance. The core customers of the computer industry valued this sustained improvement in performance, and the leading

---

[9] See Burt *et al.*, 2010, pp. 893, 894.

[10] See Burt *et al.*, 2010, p. 896.

[11] Clayton M. Christensen, 2011, *The Innovator's Dilemma: The Revolutionary Book That Will Change the Way You Do Business*, Harper Business, New York, p. 11.

firms maintained their successful standing by monitoring and responding to this customer need.

What Christensen calls a disruptive technology is one that "offered less of what customers in established markets wanted ..., [and] offered a different package of attributes valued only in emerging markets remote from, and unimportant to, the mainstream."[12] Successful firms maintain their position by responding to their core customers, but this tends to lead those firms to ignore or avoid disruptive technologies. Established customers can say what they need, but they cannot say what they don't yet know that they need. A disruptive technology builds on a new market by offering a new capability that established major markets don't (yet) need.

For example, Christensen explains (pp. 125–6) that IBM established a successful leadership role by developing, manufacturing, and selling mainframe computers to large corporations. The minicomputer was introduced in the 1970s by Digital Equipment Corporation, Data General, and others. This was a disruptive technology for which IBM's core customers had no use, but it was attractive to a small and growing market of new computer users. IBM ignored this disruptive technology in its early stages, and therefore entered the minicomputer market late, by which time its leadership role in the computer industry was already substantially reduced. Similarly, the minicomputer companies failed to exploit the disruptive potential of small desktop personal computers that were introduced by Apple, Commodore, and others. The PC appealed to customers for whom neither mainframe nor minicomputers were relevant.

The history of the computer industry illustrates that a disruptive technology identifies and nurtures a small potential market that is ignored by leaders in the industry. What Christensen calls

[12] See Christensen, 2011, p. 16.

the "innovator's dilemma" arises because disruptive technologies appeal to markets about which little is known (p. xxvi). The successful manager listens carefully to the dominant market and aggressively develops sustaining technologies that respond to the needs of the core customers. But this management strategy is paradoxical when facing a disruptive technology. The data sparsity in emerging markets, and their putatively low profit margins, discourage the successful manager from investing in disruptive technology. But this, Christensen explains, is the cause of decline in most leading firms. The innovation dilemma facing the successful manager is that a disruptive technology could either peter out or overwhelmingly supplant existing technologies—and insufficient market data and insight are available to confidently predict which it will be.

Agricultural Productivity and World Hunger

Wheat, rice, and maize are dominant components in the human diet and in animal feed and are also viewed as future sources of fuel. Global cereal yields have grown at a nearly constant yearly rate of 43 kilograms per hectare in the five decades following 1960, reaching a global average of about 3,000 kg/ha in the year 2000. These growth rates show considerable regional variation across the globe. Agricultural productivity in the UK and France rose from about 2,500 kg/ha in 1945 to about 7,000 kg/ha in the year 2000, while in the same period in Argentina, China, and India, the figure rose from about 1,000 to about 2,500 kg/ha.[13] This increasing agricultural productivity is due to innovation in many areas, including improved strains of crop plants, better

[13] R. A. Fischer, Derek Byerlee, and G. O. Edmeades, 2009, "Can Technology Deliver on the Yield Challenge to 2050?" Expert Meeting on How to feed the World in 2050, Economic and Social Development Department, Food and Agriculture Organization of the United Nations, Rome, June 24–26, 2009, pp. 1, 11.

fertilizers, more advanced farm machinery, more efficient organization of agricultural production, and more reliable irrigation. These developments can be major contributors to the elimination of poverty and malnutrition.

These agricultural innovations are clearly beneficial. However, like all innovations, they are accompanied by uncertainties that are sometimes impossible to fathom and that may result in highly undesired outcomes.

For example, many commercial food crops are cultivated in a single strain, with no variation between plants. Commercial bananas for instance are typically all biologically identical: they are clones of the same parent plant. This enables uniform quality at low cost in mass production of the bananas. However, it makes the crop highly vulnerable to attacks by pests and disease. When variability is eliminated between plants of the same crop, that crop's vulnerability is an all-or-nothing affair. For example, as Ferdman explains,[14] the Gros Michel variety of banana was virtually eliminated in the mid 1900s by the fungal Panama disease. The Cavendish variety of banana was found to resist the Panama disease, and is currently nearly the sole variety sold in the developed world. But now the Cavendish could eventually suffer the same fate as the Gros Michel, succumbing to tropical race 4, a new and more potent variety of the Panama disease. Similarly, a single variety of potato provided basic nutrition to the Irish population in the nineteenth century. The fungal Irish potato blight devastated this variety of potato in mid century, leading to malnutrition and death of hundreds of thousands of people.

These examples highlight the paradox of disease control by selecting resistant strains. On one hand, as the Cavendish replacement of the Gros Michel banana illustrates, varieties are developed to resist disease. On the other hand, selection of resistant

[14] Roberto A. Ferdman, "Bye bye, bananas", *Washington Post,* December 4, 2015.

crops can also select for resistant or more pernicious disease agents. The beneficial innovation (a more resistant food crop) can be accompanied inadvertently by a pernicious "innovation" (a more virulent disease agent).

Van den Bosch, Jeger, and Gilligan study staple food crops such as cassava, sweet potato, and plantain, where viral diseases are a major constraint, especially in developing countries. They explain that the development of improved crops needs to take into account the evolutionary response of viral disease agents.[15] How this is done requires scientific and technological innovation, and we can expect that progress will be made. The progress may result from developing the ability to quickly identify and suppress new or enhanced disease agents, or by genetic diversification (don't put all your bananas in the same genetic basket), or perhaps by some other unknown means. But just as we don't know what technologies will develop, neither do we know what new—and possibly highly pernicious—challenges will develop.

The innovation dilemma in all these cases is that a purported improvement, such as a hardier and more resistant strain of food crop, may be accompanied by a new and severe threat, such as a hardier and more pernicious agent of disease. This may vitiate or even eliminate the improvement and exacerbate the original problem.

Military Intelligence and Foresight

The killing of Osama bin Laden, founder and head of al-Qaeda, on May 2, 2011 by US Navy SEALs in Abbottabad, Pakistan, was widely acclaimed as a major accomplishment for the West over Salafi

---

[15] F. van den Bosch, M. J. Jeger, and C. A. Gilligan, 2006, "Disease control and its selection for damaging plant virus strains in vegetatively propagated staple food crops; a theoretical assessment", *Proceedings of the Royal Society, B: Biological Sciences*, vol. 274, pp. 11–18.

jihadist terror. That night US President Barack Obama declared, in an address to the nation, that "The death of bin Laden marks the most significant achievement to date in our nation's effort to defeat al-Qaeda." Al-Qaeda had openly declared war on the "far enemy": the US and other Western countries and their values. Al-Qaeda had inflicted serious injury on its enemies and had achieved wide attention and notoriety since the early 1990s, most significantly by the attacks in the US on September 11, 2001, but also in many other terrorist attacks in the Middle East, Indonesia, Kenya, and elsewhere. The death of bin Laden was rightly viewed as a turning point, but not the end of an era or an ultimate victory. President Obama acknowledged in the same address that bin Laden's "death does not mark the end of our effort. There's no doubt that al-Qaeda will continue to pursue attacks against us."

An act such as the killing of bin Laden has wide-ranging implications that are often far from clear beforehand. A major role of military intelligence is to provide background understanding and a glimpse of foresight to political or military leaders. Nonetheless, the uncertainty that pervades intelligence assessment often presents decision makers with a challenging dilemma. The dilemma need not result from a technological innovation, and can arise from an innovative and unconventional policy or strategy. The paradigm of the innovation dilemma characterizes the decision maker's problem. An example will make the point.

The West faces serious terrorist threats from numerous radical islamic organizations, including al-Qaeda, the Islamic State in Iraq and the Levant (ISIL), and many other organizations around the world. Al-Qaeda has had numerous affiliates such as Jabhat al Nusra in Syria and al-Qaeda in the Arabian Peninsula (AQAP). Likewise ISIL has branches in Libya and elsewhere. On one hand, the proliferation of these groups is a source of concern. On the other hand, these groups have been continually plagued by violent rivalry that has impeded their primary struggle against

Western interests. "Indeed, given the overall jihadist movement's bloody track record of internecine violence, the 2014 split between the Al-Qaeda core and the Islamic State in Iraq and the Levant (ISIL) is more norm than exception."[16]

The split between al-Qaeda and ISIL derived from several factors, including strategic priorities, tactics, and personality clashes. Al-Qaeda and ISIL agreed on the long-term goal of establishing a universal caliphate for instituting and enforcing the sharia (Islamic law); the strategic dispute was over timing. ISIL established this caliphate in June 2014 and declared its leader, Abu Bakr al-Baghdadi, as the caliph, while al-Qaeda insisted on deferring the restoration of the caliphate to a later date. These two organizations also disagreed on tactics. Al-Qaeda pursued a somewhat more restrained enforcement of Islamic law with the goal of winning support of Muslims under its control, while ISIL insisted on a brutal and uncompromising enforcement of sharia. Finally, a bitter personal enmity reigned between Ayman al-Zawahiri, the head of al-Qaeda, and al-Baghdadi, the leader of ISIL. These three factors—strategy, tactics, and personalities—were intertwined. In particular, the personality clash between al-Baghdadi and al-Zawahiri was a major barrier to resolving their tactical and strategic disputes.

That is our intelligence briefing. What does it imply regarding counter-terrorism strategy? The standard strategy would include persistent attempts to eliminate the leaders of these terrorist organizations. Drone technology provides a feasible means for achieving this. However, an unconventional and innovative strategy would refrain from "decapitation" of the organizational leadership, and thereby weaken both of these organizations by

---

[16] Canadian Security Intelligence Service, 2016, *Al-Qaeda, ISIL and Their Off-spring*, highlights from the workshop, February 29, 2016, CSIS National Headquarters, Ottawa, Canada, p. 23.

encouraging internecine conflict. Would decapitation be a "significant achievement" in the effort to defeat ISIL and al-Qaeda and to weaken radical Islamic terror in general? Or would the innovative strategy of leaving the leaderships intact have a greater debilitating effect on their ability to harm Western interests?

The killing of bin Laden weakened al-Qaeda, and one would expect a similar impact on ISIL or al-Qaeda if al-Baghdadi or al-Zawahiri were killed. However, the death of either leader would remove a major obstacle to the reunification of these organizations, and could thus lead, inadvertently to be sure, to a strengthening of the world jihadist threat. The radically innovative strategy of leaving the leaderships intact could have a far better outcome than the standard approach of eliminating a charismatic leader of a terrorist regime. However, not attacking a ruthless and brutal terrorist organization could reverberate globally, among both friends and foes, in ways that are very difficult to predict. The innovative strategy is less familiar and more uncertain than the standard strategy. The innovation could be much better, or far worse, than the standard decapitation strategy.

The dilemma arises if our current understanding suggests a preference for the innovative strategy of non-intervention, while our uncertainty about many aspects of the problem suggests that this option could turn out much worse than the conventional decapitation alternative. The paradigm of the innovation dilemma characterizes this situation even though new technology is not involved.

The Light Brown Apple Moth Controversy

The usual innovation dilemma results from an invention or discovery whose implications or impacts are poorly known because of its newness, and which may be either wonderful or pernicious. We now discuss a decision whose difficulty arises because we don't know if the situation we are facing is in fact new (and thus poorly understood) or old (and hence familiar). The innovation dilemma characterizes this situation as well. We will discuss a purportedly new entomological discovery with important implications for agricultural policy and practice; yet maybe it was not new and, if this is the case, it is devoid of policy implications.[17]

The Light Brown Apple Moth (*Epiphyas postvittana*) (LBAM) is a leafroller moth native to Australia, reportedly feeding on more than 250 types of plants, including many tree species, fruits, and other horticultural crops. In Australia it is generally controlled by natural predators.[18] LBAM is a non-native pest in New Zealand, where control includes releasing predators, and using insecticides and pheromones to disrupt mating behavior. It has also been found in New Caledonia, Hawaii, and England.[19]

LBAM was first detected in the continental United States in Berkeley, California in 2007. The species was not subject to detection efforts, but evaluation of trapping evidence indicated it might have been present in 2006. In order to prevent damage to California crops and to avoid the imposition of trade bans by other jurisdictions, the US Department of Agriculture (USDA) and State officials launched an eradication and quarantine program

[17] Yakov Ben-Haim, Craig D. Osteen, and L. Joe Moffitt, 2013, "Policy dilemma of innovation: An info-gap approach", *Ecological Economics,* 85: 130–8. Science for Environment Policy, 2013, "Choosing between established and innovative policy measures: controlling invasive species", European Commission DG, Environment News Alert Service, edited by SCU, The University of the West of England, Bristol. http://ec.europa.eu/environment/integration/research/newsalert/pdf/325na2.pdf.

[18] Daniel Harder and Jeff Rosendale, 2008. "Integrated Pest Management Practices for the Light Brown Apple Moth in New Zealand: Implications for California". March 6, 2008, pp. 14. http://www.lbamspray.com/00_Documents/2008/HarderNZReportFINAL.pdf, accessed July 24, 2011.

[19] United States Department of Agriculture, 2011, "Light Brown Apple Moth", http://www.aphis.usda.gov/plant_health/plant_pest_info/lba_moth/background.shtml, accessed July 2, 2011.

that involved delimiting surveys, aerial pesticide applications (LBAM pheromones), and development of integrated pest management methods, with eradication anticipated by 2011 followed by control maintenance activities.[20] The program allowed Canada, Mexico, and other US states to relax trade restrictions and accept LBAM-host crops from non-infested California counties. In late 2007 it was estimated that crop losses could reach $2.6 billion annually if LBAM was not controlled and entered the San Joaquin Valley. In 2007 and 2008, the USDA allocated about $90 million in emergency funding to the LBAM program.

At the same time, several University of California entomologists disagreed with the official assessment, maintaining that the dispersion of LBAM over hundreds of miles indicated that it had been present in California for decades, could not be eradicated, and was causing no crop damage.[21] Public distaste for aerial spraying in many affluent areas led to multiple petitions filed with the Secretary of Agriculture to declassify LBAM as an "actionable" pest.[22] Legal challenges led to a state court ruling in Spring 2008 suspending the aerial spraying program.[23]

[20] United States Department of Agriculture, 2008, "USDA Budget Explanatory Notes for the Committee on Agriculture for FY 2009, FY 2010, and FY 2011".

[21] J. R. Carey, F. G. Zalom, and B. D. Hammock, 2008, "Concerns with the eradication program against the light brown apple moth in California". Personal communication to E. Schafer on May 28, 2008, from J. Carey, F. G. Zalom, and B. D. Hammock. Ingfei Chen, 2010, "From medfly to moth: Raising the buzz of dissent", *Science*, 327 (5962), 134–6, January 8, 2010.

[22] Daniel Harder, Ken Kimes, Roy Upton, and Lynette Casper, 2009, "Light Brown Apple Moth (LBAM) Eradication Program: Formal Petition to Recalssify LBAM as a Non-quarantinable Pest: Summary of Findings". January 7, 2009. http://www.lbamspray.com/Reports/ReclassificationPetitionSummaryRU.pdf, accessed July 3, 2011.

[23] United States Department of Agriculture, 2010, "Animal and Plant Health Inspection Service Draft Response to Petitions for the Reclassification of Light Brown Apple Moth [*Epiphyas postvittana* (Walker)] as a Non-Quarantine Pest", Revision January 14, 2010. http://www.aphis.usda.gov/plant_health/plant_pest_info/lba_moth/downloads/draft_lbam_petition_response-10.pdf.

The controversy over the LBAM eradication program presented policy officials with an innovation dilemma. The eventual economic impact of the LBAM, and even whether it was a newly invasive species, are both highly uncertain, and different credible experts hold widely different opinions. We will refer to three groups of experts. *Optimists* claim that the species is harmless and intervention is not needed and that in fact, nothing is new so there is no need for change. *Activists* claim that the species was recently introduced and is harmful but can be eradicated or at least substantially contained by actions costing much less than the potential loss. *Pessimists* agree that the species is harmful but claim that intervention can have no useful impact. All the claims are uncertain. The activist claim is the most uncertain and entails the possibility of collateral damage so that the total damage could exceed the pessimist prediction.

If the activists are right, then action must be taken (the optimists err) and the outcome will be better than the pessimists' prediction. But if the pessimists are right, then activist intervention cannot help and would make things even worse than the pessimist prediction. In other words, activism could be either much better, or much worse, than pessimism. The innovation dilemma is that we don't know if the situation is new and volatile, requiring a new policy, or if the situation is long-standing and stable. The severe uncertainty prevents us from knowing which is correct. The challenge facing the policymaker has the structure of an innovation dilemma even though no technological innovation is involved.

The Habit of Open-Mindedness

Habits are important and valuable, though they may also be detrimental. The habit of open-mindedness is a case in point because it constitutes an innovation dilemma.

David Hume explained that we believe by habit that logs will burn, stones will fall, and endless other past patterns will recur.[24] No experiment can prove the future recurrence of past events. An experiment belongs to the future only until it is implemented; once completed, it becomes part of the past. In order for past experiments to prove that past phenomena will also be observed in the future, we must assume that the past will recur in the future. That's as circular as it gets.

But without the habit of believing that past patterns will recur, we would be incapacitated and ineffectual (and probably reduced to moping and sobbing). Who would dare climb stairs, or fly planes, or eat bread and drink wine, without the belief that, like in the past, the stairs will bear our weight, the wings will carry us aloft, and the bread and wine will nourish our body and soul.[25] Without such habits we would become jittering jellies of indecision in the face of the unknown.

But you can't just pull a habit out of a hat. We spend great effort instilling good habits in our children: to brush their teeth, tell the truth, and not pick on their little sister even if she deserves it.

As we get older we begin to worry that our habits are becoming frozen, stodgy, closed-minded, and constraining. Younger folks smile at our rigid ways, and try to loosen us up to the new wonders of the world, be they technological, culinary, or musical. Changing your habits, or staying young when you aren't, isn't ever easy. Without habits we're lost in an unknowable world.

And yet, openness to new ideas, tastes, sounds, and other experiences of many sorts can itself be a habit, and perhaps a good one. It is the habit of testing the unknown, of acknowledging the great

[24] David Hume, 1748, *An Enquiry Concerning Human Understanding*, section 12, part III.

[25] Psalm 104: 15.

gap between what we *do* know and what we *can* know. That gap is an invitation to growth and awe, as well as to fear and danger.

The habit of openness to change is not a contradiction. It is simply a recognition that habits are a response to the unknown. Not everything changes all the time (or so we're in the habit of thinking), but some things *are* new under the sun (as newspapers and Nobel prize committees periodically remind us).

The habit of open-mindedness is paradoxical. On one hand habits are, by definition, conservative. We repeat past practices by habit. On the other hand, the habit of open-mindedness can bring us to abandon our customs and former habits, perhaps even abandoning the habit of open-mindedness.

This paradox is an innovation dilemma. Openness to new objects, ideas, tastes, and behaviors is potentially beneficial but possibly dangerous, where the uncertainty results from the newness. Habits, including the habit of open-mindedness, are a good thing even though we can never know for sure how good or bad they really are.

3

Uncertainty, Ignorance, Surprise —The Endless Frontier

"'Now I'll give *you* something to believe. I'm just one hundred and one, five months and a day.'

"'I can't believe *that*!' said Alice.

"'Can't you?' the Queen said in a pitying tone. 'Try again: draw a long breath, and shut your eyes.'

Alice laughed. 'There's no use trying,' she said: 'one *can't* believe impossible things.'

"'I daresay you haven't had much practice,' said the Queen. 'When I was your age, I always did it for half-an-hour a day. Why, sometimes I've believed as many as six impossible things before breakfast.'"[1]

Some things happen by chance. Chance has diverse interpretations, but probability theory is relevant if there is some regularity of events. Probability also has limitations, especially when considering very rare events that we can only vaguely imagine. There are endless unknown possibilities, and the totality of rare events

---

[1] Lewis Carroll (Charles Lutwidge Dodgson), 1871, *Through the Looking-Glass*, chapter 5, Project Gutenberg, Salt Lake City, UT, last updated October 20, 2015.

The Dilemmas of Wonderland: Decisions in the Age of Innovation. Yakov Ben-Haim.
© Yakov Ben-Haim 2018. Published in 2018 by Oxford University Press.
DOI: 10.1093/oso/9780198822233.001.0001

is massive. Hence it is hard to estimate probabilities of events that are individually rare but collectively not so rare. The rareness illusion is the impression of rareness arising from ignorance of the unknown. The rareness illusion results from our inability to assess probabilities of rare events. Non-probabilistic Knightian uncertainty extends our understanding of the unknown, and Shackle–Popper indeterminism provides a logical foundation. We illustrate these ideas by discussing the vagueness of human language, the impossibility of one unified theory of uncertainty, and the paradox that science is possible because science could someday come to an end.

We Call It Chance

Nothing is as ancient as the feeling that some things are beyond our control. Many things happen—or don't happen—regardless of our plans or wishes, sometimes for worse but sometimes for better. "Fortune brings in some boats that are not steer'd."[2]

Tradition has it that the poet Robert Burns was plowing his field one autumn when he accidently destroyed the burrow of a mouse. With winter coming on, this would surely mean the death of the mouse. Burns wrote:[3]

The best-laid schemes o' mice an' men
Gang aft agley,[4]
An' lea'e us nought but grief an' pain,
For promis'd joy!

[2] William Shakespeare, *Cymbeline*, Act I, scene iii.
[3] Robert Burns, 1785, "To A Mouse", *The Kilmarnock Edition of the Poetical Works of Robert Burns*, William Scott Douglas, ed., Scottish *Daily Express*, Glasgow, 1938.
[4] "Go often askew" is the English translation of this Scots phrase.

Misfortune happens to both mice and men, but ignorance and surprise can be created and exploited. The book of *Proverbs* advises the use of tricks and ruses in war,[5] and Sun Tzu agrees:[6]

> All warfare is based on deception. Hence, when we are able to attack, we must seem unable; when using our forces, we must appear inactive; when we are near, we must make the enemy believe we are far away; when far away, we must make him believe we are near.

By deceiving the enemy one can defeat him by surprise.

The churning of chance can also solve problems and soothe feelings, as the following example shows. The ancient temple of the Jews in Jerusalem, first built by Solomon in the tenth century BCE, was the site of many rituals and sacrifices. The sons of Aaron were designated as servants of the Lord, *cohanim*:[7] functionaries in performing the rites of the temple. These sacred tasks were much coveted by the *cohanim*, and the allocation of tasks was originally done each morning by having the day's candidates race on foot up a temple ramp, the winner getting the first task, and so on.

This turned out to be dangerous. It once happened that one candidate in the race pushed his fellow contestant, who fell off the ramp and broke his leg. The legal court then discontinued these allocation races and instituted a lottery instead. The day's cohort of *cohanim* formed a circle around the convening *cohen* who picked a number much larger than the number of *cohanim* present. Each *cohen* then stuck out one or several fingers which the convenor counted, going around and around the circle until reaching the

---

[5] 24:6. The book of *Proverbs* may derive from an Egyptian source far pre-dating Sun Tzu. See E. W. Heaton, 1974, *Solomon's New Men: The Emergence of Ancient Israel as a National State*, Thames & Hudson, London.

[6] Sun Tzu, *The Art of War*, trans. by Samuel B. Griffith, 1963, Oxford University Press, chapter 1.

[7] *Cohanim* is the plural form; the singular is *cohen*.

number he had picked. The *cohen* on whom the counting ceased won the first allocation. The procedure was repeated until all the rituals had been allocated. This procedure made a fair and equal allocation, prevented favoritism or jealousy, and avoided the dangers of the race up the ramp.[8]

Was the outcome of the lottery a result of fate, or celestial circumstance, or the will of the Lord? We don't really know how the ancient *cohanim* understood this lottery. We call it chance.

From Chance to Probability

Is chance impenetrable? Or can one see patterns, make predictions, and understand the causes and processes that produce chance occurrences?

The modern answers to these questions began to emerge early in the seventeenth century with the study of games of chance, leading eventually to the mathematical theory of probability.

Consider, for instance, the rolling of dice. Each die is a small, uniformly balanced cube with a different number of dots, from one to six, on each face. If you shake and roll a die it will come to rest with one face upwards. Repeating this many times, you will sometimes get several "twos" in a row, then a "six" followed by a "one", and so on. Each outcome occurs by chance, but a pattern emerges over a long number of throws: each face occurs nearly the same number of times, though in no particular order.

We can summarize this by saying that the **empirically estimated frequency** of occurrence is about one-sixth for each face. The leap of abstraction from this sort of observation is the idea of probability: each face seems equally likely to occur. Turning this observation into mathematics, we say that each face has a

[8] *Yoma*, chapter 2, verses 1–2. See the talmudic commentaries of Rabbi Ovadia of Bartenura as well as *Melechet Shlomo.*

probability of occurrence equal to about one-sixth. Similarly, the frequency of occurrence of osteoporosis is about 20 percent in Caucasian women over the age of 50. That is, a woman in this category has probability of about one-fifth of developing osteoporosis. In January, about 48 percent of the days are rainy in Haifa, while only 28 percent are rainy in Sydney. In other words, the probability of a rainy January day is about one-half in Haifa and about one-fourth in Sydney.

Going back to the rolling of dice, we notice that the die is symmetric and uniformly balanced, even before we roll it at all. From this we infer that each face has the same chance of turning up (like each *cohen* had the same chance of being chosen). That is, we make a **logical deduction** of the probability of each face turning up. This is a prediction—not an observation—of a general pattern describing the chance occurrences of repeatedly rolling a die. The logical deduction predicts that, as the number of rolls increases, the fractions of occurrence converge ever more closely to their theoretical probabilistic limit. The prediction is that the frequency of each face will be approximately one-sixth because the logical probability of each face turning up is exactly one-sixth.

We can also use the idea of probability to understand the causes or mechanisms that generate chancy events. For example, if we observe a preponderance of "twos" in a long sequence of outcomes of rolling a die, then we conclude that that particular die is *not* uniformly balanced. We may even infer something about the owner of that die.

We can also use the idea of probability in situations where neither logical deduction nor systematic empirical observation are possible. One can make a **personal judgment** of the likelihood of an event. For example, Stalin's military advisers in 1941 claimed that an imminent German invasion of the Soviet Union was very likely. The advisers had reconnaissance evidence, captured documents, and more.[9] This is neither a logical deduction, nor an empirical estimate based on quantitative observations of a controlled group of subjects. This is expert judgment of likelihood based on verbal and documentary evidence and experience in related affairs.

The theory of probability enables prediction of outcomes and understanding of causes in a vast array of situations. Probability is versatile because it can be interpreted in diverse ways, reflected by the three different interpretations of probability that we have discussed: empirical, logical, and personal. Probability is particularly powerful when linked to statistical tools of inference and estimation.

The Rareness Illusion and the Limits of Probability

Probability does, however, have its limitations. This is especially evident in situations where we just can't know the probabilities of the relevant events. For instance, it is often quite difficult to know or estimate the probabilities of very rare events, ones we've never witnessed or can only vaguely imagine. Each rare event is, of course, rare and elusive. However, the realm of unknown possibilities is so vast that the totality of all rare events is massive. This means that it is hard to know or estimate the probabilities of individual rare events. An example will illustrate the idea.

Suppose you decide to spend the summer roaming the world in search of the ten lost tribes of Israel, exiled from Samaria by the Assyrians 2,700 years ago.[10] Or perhaps you'd like to search for Prester John, the virtuous ruler of a kingdom lost in the Orient? Or would you rather trace the gold-laden kingdom of Ophir?[11] Or do you prefer the excitement of tracking the Amazons, that

[9] Richard Overy, 2006, *Why the Allies Won*, 2nd edn., London, Pimlico, p. 80.

[10] 2 Kings 17:6.

[11] 1 Kings 9:28.

nation of female warriors? Or perhaps the supposedly mythical naval power mentioned by Plato,[12] operating from the island of Atlantis? Or how about unicorns, or the fountain of eternal youth? The unknown is so vast that the possibilities are endless.

Maybe you don't believe in unicorns. But Plato might really have believed the history of Atlantis reported by Timeaus and Critias. The conquest of Israel is known from Assyrian archeology and from the Bible. Does the fact that you've never seen a Reubenite or a Naphtalite (or a unicorn) mean they don't exist?

It is true that when something really does not exist, one might spend a long time futilely looking for it. Many people have spent enormous energy searching for lost tribes, lost gold, and lost kingdoms. Why is it so difficult to decide that what you're looking for really isn't there? The answer is that we have almost no way of knowing the probability that it's really not there. And the reason for that, ironically, is that the world has endless possibilities for discovery and surprise.

Let's consider some real-life searches. Civil war began in Libya in February 2011. Tripoli, the capital, fell in August. How long should you (or the Libyans) have looked for Muammar Gaddafi, the fallen dictator, before giving up? After all, other dictators have seemed to vanish without a trace. If he's not in Tripoli, maybe he's in Bani Walid, or Algeria, or Timbuktu? How do you decide he cannot be found? Maybe he was pulverized by a NATO bomb.[13] Here's another example. It's urgent to find the suicide bomber in the crowded bus station before it's too late—if he's really there. Or, you'd like to discover a cure for AIDS, or a method to halt the rising global temperature, or a golden investment opportunity in an emerging market, or a proof of the parallel postulate of Euclidean geometry. How long should you continue the search?

[12] Plato's dialogues of *Timeaus* and *Critias.*
[13] Gaddafi was captured and killed in Sirte, Libya, on October 20, 2011.

Let's focus our question. Suppose you are looking for something, and so far you have only "negative" evidence: it's not here, it's not there, it's not anywhere you've looked. Why is it so difficult to decide, conclusively and confidently, that it simply does not exist?

Answers can be found in several domains.

Psychology provides some answers. People can be very goal-oriented, stubborn, and persistent. Marco Polo didn't get to China on a ten-hour plane flight. The round trip took him 24 years, he was on the road for at least 7 years, and he didn't travel business class. Some people simply refuse to give up.

Ideology is a very strong motivator. When people believe something strongly, it is easy for them to ignore evidence to the contrary. Furthermore, for some people, the search itself is valued more than the putative goal.

The answer that I will focus on is found by contemplating the endless unknown. The unknown is so vast, so unstructured, so, well..., *unknown*, that we cannot calibrate our negative evidence in terms of the probability that whatever we're looking for really isn't there.

I'll tell a true story. I was born in the US and my wife was born in Israel, but our life-paths crossed, so to speak, before we were born. She had a friend whose father was from Europe and lived for a while—before the friend was born—with a cousin of his in my home town. This cousin was—years later—my third grade teacher. My schoolteacher was my future wife's friend's father's cousin.

Amazing coincidence. This convoluted sequence of events is certainly rare. How many people can tell the very same story? But wait a minute. This convoluted string of events could have evolved in many many different ways, each of which would have been an equally amazing coincidence. The number of similar possible paths is namelessly enormous, uncountably humongous.

In other words, potential "rare" events are very numerous. Now that sounds like a contradiction (we're getting close to some of Zeno's paradoxes, and Aristotle thought Zeno was crazy).

But it is not a contradiction; it is only a "rareness illusion" (something like an optical illusion). The specific event sequence in my story is unique, which is the ultimate rarity. We view this sequence as an amazing coincidence because we cannot assess the number of similar sequences. This means that we cannot estimate the probability of a similar sequence occurring. But actually, surprising strings of events occur not infrequently because the number of possible surprising strings is so unimaginably vast. *The rareness illusion is the impression of rareness arising from our ignorance of the vast unknown.* The rareness illusion results from our inability to assess probabilities of rare events. We are ignorant of the unknown because, by definition, we do not know what is unknown. The unknown is vast because the world is so rich in possibilities. We can't estimate the probabilities of rare events because we can't count or calibrate them.

The rareness illusion is a false impression, a mistake: a sense of surprise resulting from our ignorance of the vast range of possible similar surprises. The rareness illusion leads people to wrongly goggle at strings of events—individually rare in themselves—even though "rare events" are numerous and "amazing coincidences" occur not infrequently.

Recognition of the rareness illusion is the key to understanding why it is so difficult to confidently decide, based on negative evidence, that what you're looking for simply does not exist.

One might be inclined to reason as follows. If you're looking for something, then look *very* thoroughly, and if you don't find it, then it's not there. That is usually sound and sensible advice, and "looking thoroughly" will often lead to discovery.

However, the number of ways that we could overlook something that really *is* there is enormous. It is thus very difficult to confidently conclude that the search was thorough and that the object cannot be found. Take the case of your missing house keys. They dropped from your pocket in the car, or on the sidewalk and somebody picked them up, or you left them in the lock when you left the house, or.... Familiarity with the rareness illusion makes it very difficult to decide that you have searched thoroughly. If you think that the only contingencies not yet explored are too exotic to be relevant (a raven snatched them while you were daydreaming about that enchanting new employee), then think again, because you've been blinded by the rareness illusion. The number of such possibilities is so vastly unfathomable that you cannot confidently say that all of them are collectively negligible. Recognition of the rareness illusion prevents you from confidently concluding that what you are seeking simply does not exist.

The rareness illusion is one example where probability is not useful, and other methods are called for. Before we are ready for that (in Chapter 5), we need to broaden our understanding of uncertainty, and reconnect to the innovation dilemma.

Knightian Uncertainty

In the previous chapter we examined a number of innovation dilemmas arising from the uncertainty that surrounds a new invention, discovery, or course of action. In each case, one option (the innovation) purports to be much better than a more standard option, but the innovative option is more uncertain and less familiar and thus may be much worse; hence the dilemma. Digital devices provide many promising attractions for young readers, but we don't know their long-term impact on cognitive development, parent–child relations, and other issues (p. 7). Innovative military hardware is crucial in establishing dominance in battle, but development of such technology may require far greater resources of time, money, and expertise than initially anticipated, and success is still not guaranteed (p. 9). The management of psychiatric conditions can be greatly enhanced by medically supervised use of drugs, and new drugs are continually being developed. However, the impact of powerful new drugs on the delicate psychological and physiological balance of a specific individual is very difficult to anticipate (p. 10). After several more examples we found that we are so used to surprises that we sometimes see them where they aren't. Or, more precisely, we may be very uncertain if the situation we face is fundamentally new or not, as illustrated by the Light Brown Apple Moth controversy (p. 20).

All of these situations, and others like them, are highly uncertain, indeterminate, vulnerable to surprise, and so on. However, concepts of probability are not particularly useful in characterizing and managing the info-gaps in these situations, like the rareness illusion discussed in the previous section. Rather, the concept of Knightian uncertainty is central, as we now explain.

We should not interpret the uncertainty around an innovation dilemma in the same way that we interpret the assertion that "The probability of 'snake eyes' on the role of two dice is 1 out of 36." This assertion is a **logical deduction** from the assumption of equal probabilities for each of six outcomes for each die, and independence of the two dice. Of the thirty-six possible outcomes—all equally likely—only one is snake eyes, hence the assertion. An innovation often has attributes that are incompletely understood, in which case logical deduction of probabilities is not possible.

Nor should we interpret those descriptions of uncertainty in the same way that we understand the statement that the incidence of tuberculosis in Europe is 20 per 100,000 people, or that 29% of professional baseball starting pitchers are left-handed, while about 10% of the general population are left-handed. These are **empirical estimates** based on measurements and observations. We usually lack extensive data about innovations and their impacts, precluding empirical estimates of frequencies.

Finally, statements about the uncertainty surrounding an innovation dilemma should not be understood as we would understand considered personal judgment of likelihood. For example, Keynes predicted that the severe economic penalties extracted from Germany in the Versailles treaty after World War I would very likely lead to war in Europe within a generation.[14] This is neither a logical deduction nor an empirical estimate based on measurements. This is expert **personal judgment** of likelihood based on insight and experience. Innovations are new, so relevant expertise and experience are usually limited.

These three types of probability statements—logical, empirical, and personal—are diverse, but they all share the conceptual and axiomatic framework of probability theory.

Probabilistic judgments are very different from statements of uncertainty associated with an innovation dilemma. An essential challenge of an innovation dilemma is that vast amounts of relevant evidence, experience, and understanding are lacking, and that surprise and ignorance are prevalent. It is often infeasible to make reliable or justifiable probabilistic statements about the info-gaps surrounding an innovation dilemma. For example, the logical deduction about the outcome of rolling six-sided balanced dice cannot be extended to rolling dice of unknown geometry and composition. Empirical statements of frequency cannot be made about populations or circumstances that have never been observed or do not yet exist, such as the rate of spread of a new and unfamiliar disease. Personal judgments are of diminutive value in the absence of evidence or experience with analogous situations.

[14] John Maynard Keynes, 1920, *The Economic Consequences of the Peace*, Harcourt, Brace and Howe, New York.

Frank Knight, an economist, recognized this fundamental distinction between probability and what he called "true uncertainty".[15] In many situations one knows the probability of uncertain events, such as natural disasters or ordinary economic fluctuations. One then faces what Knight called "risk", and it is possible to buy and sell insurance against this risk in a systematic and sustainable way. When likelihoods of all possible outcomes are known, one can calculate the frequency of adverse (and beneficial) events and this underlies a viable insurance market.

Knight also recognized that some uncertainties cannot be known probabilistically, for example surprising innovations by one's competitors, or sudden change in public preference for one product over another. To this we would add unanticipated social upheaval, surprising impediments in developing a new product, and unknown interactions between systems that have never been combined before. One lacks probabilistic information about these situations and Knight referred to "true uncertainty" as distinct from probabilistic risk. Knight's concept of true uncertainty—as distinct from probability—has come to be known as Knightian uncertainty.

Shackle–Popper Indeterminism

Knightian uncertainty is central to understanding innovation dilemmas because it results from the human capacity to invent or discover and then to alter behavior as a result. We now discuss a generic mechanism that is responsible for the prevalence of Knightian uncertainty in human affairs.

---

[15] Frank H. Knight, 1921, *Risk, Uncertainty and Profit*, Hart, Schaffner and Marx, reissued by Harper Torchbooks, New York, 1965, pp. 46, 120, 231–2. See also Frank H. Knight, 1933, *The Economic Organization*, reissued by Harper Torchbooks, New York, 1951, p. 120.

The power of probability theory derives from its ability to capture the patterns that underlie random processes as diverse as games of chance, electrical currents, stock price fluctuations, and solar flares. Much human behavior is fruitfully characterized with probability, even when the mechanisms are totally unknown. An early example was demonstrated in the late eighteenth and early nineteenth centuries. Each year the central post office of Paris accumulated a large number of letters that were undeliverable for various reasons: unknown addressee, nonexistent or illegible address, and so on. The precise number of so-called dead letters varied from year to year, but the average was remarkably constant over the years.[16] So simple a probabilistic idea as calculating the average revealed something systematic about the complicated underlying process.

Human affairs are, however, often characterized by invention, innovation, discovery, and surprise. Our habits and preconceptions change, as do our opportunities, ideals, tools, and institutions. Change is sometimes rapid and sometimes slow, but human behavior is altered fundamentally, irrevocably, and unpredictably, by the fruits of our creative potential. The world was a different place—people behaved differently—after the invention of the printing press or the discovery of America. Separately, and in different ways, the economist G. L. S. Shackle[17] and the philosopher Karl Popper[18] developed an idea

[16] Pierre-Simon, Marquis de Laplace, *A Philosophical Essay on Probabilities*, Trans. by F. W. Truscott and F. L. Emory, 1951, Dover Publications, New York, p. 62. See also Theodore M. Porter, 1986, *The Rise of Statistical Thinking: 1820–1900*, Princeton University Press, Princeton, New Jersey, p. 51.

[17] G. L. S. Shackle, 1972, *Epistemics and Economics: A Critique of Economic Doctrines*, Transaction Publishers, 1992, originally published by Cambridge University Press, New Brunswick and London, pp. 3–4, 156, 239, 401–2.

[18] Karl Popper, 1982, *The Open Universe: An Argument for Indeterminism.* From the Postscript to *The Logic of Scientific Discovery*, Routledge, London and New York, pp. 80–81, 109.

of indeterminism in human affairs that we will now describe.[19] Shackle–Popper indeterminism is important for our discussion because it explains the prevalence of Knightian uncertainty.

The basic claim made, independently, by Shackle and by Popper is that the behavior of intelligent systems displays an element of unstructured and unpredictable indeterminism. This claim rests on two basic concepts: intelligence and learning.

By *intelligence* is meant: behavior is influenced by knowledge; what a person or group does is influenced by what they know. This is surely characteristic of humans individually and of groups and of society at large. If you know it will rain, you will take an umbrella; if you hear that the market crashed this morning, you won't sell your stocks this afternoon.

By *learning* is meant a process of discovery or invention: finding out today what was unknown yesterday. The invention of writing and the discovery of the polio vaccine profoundly impacted our behavior. Even commonplace learning can be important. One economically important example of learning is what Keynes referred to as hearing "the news" about daily events, even though these events may be mundane.[20]

Finally, *indeterminism* arises as follows: because tomorrow's discovery or invention is by definition unknown today, tomorrow's behavior is not predictable today, at least not in its entirety. Given the richness of future discovery, (or its corollary, the richness of our current ignorance), the indeterminism of future behavior is broad, deep, and unstructured. The laws of behavior of an intelligent learning system will evolve in time as the agents make

[19] Yakov Ben-Haim, 2007, Peirce, Haack and Info-gaps, in *Susan Haack, A Lady of Distinctions: The Philosopher Responds to Her Critics*, edited by Cornelis de Waal, Prometheus Books, Amherst, New York.

[20] John Maynard Keynes, 1936, *The General Theory of Employment, Interest, and Money*, Harcourt Brace & World, reissued by Prometheus Books, Amherst, New York, 1997, pp. 198, 199, 204.

discoveries and inventions. These laws cannot be entirely known ahead of time. Indeed, they don't exist in their mature form until they emerge, because by definition discoveries cannot be predicted and the laws of behavior depend, in part, on discoveries to be made.

The implication of Shackle–Popper indeterminism for all human activity is that the power of scientific understanding and of its predictive models is unavoidably limited to some extent by ongoing invention and discovery. We live in an "open universe" (to use Popper's phrase) in which new things will be discovered and invented. This refers to great inventions or discoveries (the domestication of animals or the personal computer) as much as to more modest inventions (the stirrup or automatic transmission) or truly humble ones (meeting new neighbors or discovering the joys of coffee). Our understanding of the rules by which a system evolves will change in unpredictable ways and at unpredictable instants. Models and theories can and will improve, but the conception that they converge on a stable and universal truth is fundamentally incorrect in the open universe of creative human activity.

Shackle–Popper indeterminism also implies that the structured conception of uncertainty upon which probability theory is based cannot be extended to the inventions and discoveries that cause indeterminism in human affairs. Probability theory depends on knowing the space of possible events, whether simple or complicated: six outcomes of rolling a die, or continuous fluctuations of wave-shapes at the beach. We cannot know what is not yet discovered, so the event-space underlying Shackle–Popper indeterminism is in part hidden from us. Furthermore, future likelihoods in human affairs are themselves indeterminate because new discoveries can alter them. Probability is powerful, but not all-powerful, in dealing with uncertainty. Shackle–Popper indeterminism is not probabilistic, and it explains the origin of non-probabilistic Knightian uncertainty.

Baseball and Linguistic Uncertainty

We've discussed the idea of Knightian uncertainty, its distinction from probability, and the concept of Shackle–Popper indeterminism. We now consider a more mundane example of deep uncertainty: the vagueness or ambiguity of human language.

A sports commentator, in discussing the art of pitching, once said that throwing a baseball is like shooting a shotgun. You get a spray. As a pitcher, you have to know your spray. You learn to control it, but you know that it is there. The ball won't always go where you want it. And furthermore, where you want the ball depends on the batter's style and strategy, which vary from pitch to pitch for every batter.

Baseball pitchers must manage uncertainty, and it is not enough to reduce it and hope for the best. Suppose you want to throw a strike. It's not a good strategy to aim directly at, say, the lower outside corner of the strike zone, because of the spray of the ball's path and because the batter's stance can shift. Especially if the spray is skewed down and out, you'll want to move up and in a bit.

This is all very similar to the ambiguity of human speech when we pitch words at each other. Words don't have precise meanings; meanings spread out like the pitcher's spray. If we want to communicate precisely we need to be aware of this uncertainty, and manage it, taking account of the listener's propensities.

Take the word "liberal" as it is used in political discussion. For many decades, liberals in the United States and elsewhere have tended to support high taxes to provide generous welfare, public medical insurance, low-cost housing, and other public goods. They advocate liberal (meaning magnanimous or abundant) government involvement for the citizens' benefit.

A liberal might also be someone who is open-minded and tolerant; who is not strict in applying rules to other people, or even to him or herself. Such a person might be called "liberal" (meaning advocating individual rights) for opposing extensive government involvement in private decisions. For instance, liberals (in this second sense) might oppose high taxes since they reduce individuals' ability to make independent choices. As another example, John Stuart Mill opposed laws that restricted the rights of women to work (at night, for instance), even though these laws were intended to promote the welfare of women. Women, insisted Mill, are intelligent adults and can judge for themselves what is good for them.[21]

Returning to the first meaning of "liberal" mentioned above, people of that strain may support restrictions of trade to countries that ignore the health and safety of workers. The second type of liberal might tend to support unrestricted trade.

Sending out words and pitching baseballs are both like shooting a shotgun: meanings (and baseballs) spray out. You must know what meaning you wish to convey, and what other meanings the word can have. The choice of the word, and the crafting of its context, must manage the uncertainty of where the word will land in the listener's mind.

Let's go back to baseball. If there were no uncertainty in the pitcher's pitch and the batter's swing, then baseball would be a dreadfully boring game. If the batter knows exactly where and when the ball will arrive, and can completely control the bat, then every swing will be a homer. Or conversely, if the pitcher always knows exactly how the batter will swing, and if each throw is perfectly controlled, then every batter will strike out. But which is it? Whose certainty dominates? The batter's or the pitcher's? It can't be both. There is some deep philosophical problem here. Clearly there cannot be complete certainty in a

---

[21] John Stuart Mill, 1869, *The Subjection of Women*, Longmans, Green, Reader & Dyer, London.

world that has some element of free will, or surprise, or discovery. Uncertainty—which makes baseball and life interesting—is inevitable in the human world.

How does this carry over to human speech? It is said of the Wright brothers that they thought so synergistically that one brother could finish an idea or sentence begun by the other. If there is no uncertainty in what I am going to say, then you will be bored with my conversation, or at least you won't learn anything from me. It is because you *don't* know what I mean by, for instance, "robustness", that what I have to say on this topic is enlightening (and maybe interesting). And it is because you disagree with me about what robustness means (and you tell me so), that I can perhaps extend my own understanding.

So, uncertainty is inevitable in a world that is rich and varied enough to have surprise and free will. Furthermore, this uncertainty leads—often through speech—to a process of discovery and new understanding. Uncertainty, and the use of language, leads to discovery, and discovery induces new uncertainties.

Grand Unified Theory of Uncertainty???

I am referring to uncertainty as though it were some homogeneous and coherent phenomenon, like English cooking, the evolution of galaxies, or the history of powered locomotion. Could one write a fairly definitive book on uncertainty? Can one realistically strive towards one grand unified theory of uncertainty, like the physical scientists who seek a Grand Unified Theory of the universe?

We've had our cerebral cortex for several tens of thousands of years. We've lived in more or less sedentary settlements and produced excess food for seven or eight thousand years. We've written down our thoughts for roughly five thousand years.

And science? The ancient Greeks had some,[22] but science and its systematic application are overwhelmingly a European invention of the past 500 years. We can be proud of our accomplishments (quantum theory, polio vaccines, powered machines), and we should worry about our destructive capabilities (atomic, biological, and chemical weapons). But it is quite plausible, as Koestler suggests, that we've only just begun to discover our cerebral capabilities.[23] It is more than just plausible that the mysteries of the universe are still largely hidden from us. As evidence, consider the fact that the main theories of physics— general relativity, quantum mechanics, statistical mechanics, thermodynamics—are still not unified. And it goes without saying that the consilient unity of science[24] is still far from us. Nonetheless, the aspiration of the scientists is to construct one Grand Unified Theory of the universe.

A theory, in order to be scientific, must exclude something. A scientific theory makes statements such as "This happens; that doesn't happen." Karl Popper explained that a scientific theory must contain statements that are at risk of being wrong, statements that could be falsified.[25] Deborah Mayo demonstrated how science grows by discovering and recovering from error.[26] The foremost method of discovering error in a theory is by uncovering a contradiction, either between observation and prediction, or between two implications of the theory.

[22] Marshall Clagett, 1955, *Greek Science in Antiquity*, Collier Books, New York. Lucio Russo, 2003, *The Forgotten Revolution: How Science was Born in 300 BC and Why It Had to Be Reborn*, with the collaboration of the translator, Silvio Levy, Springer, Berlin.

[23] Arthur Koestler, 1967, *The Ghost in the Machine*, Hutchinson of London.

[24] Edward O. Wilson, 1998, *Consilience: The Unity of Knowledge*, Random House, New York.

[25] Karl R. Popper, 1968, *The Logic of Scientific Discovery*, Harper & Row, New York.

[26] Deborah G. Mayo, 1996, *Error and the Growth of Experimental Knowledge*, University of Chicago Press, Chicago.

Science has evolved by developing diverse and mutually inconsistent theories, and is still far from unification. What holds for
science in general, holds also for the study of uncertainty. The
ancient Greeks invented the axiomatic method and used it in the
study of mathematics but did not concern themselves with the
mathematics of uncertainty.[27] Some medieval thinkers explored
the mathematics of uncertainty,[28] but it wasn't until around 1600
that serious thought was directed to the systematic study of
uncertainty,[29] and statistics as a separate and mature discipline
emerged only in the nineteenth century.[30] The twentieth century saw a florescence of uncertainty models. Lukaczewicz discovered three-valued logic in 1917,[31] and in 1965 Zadeh introduced his work on fuzzy logic.[32] In between, Wald formulated a
modern version of min-max in 1945.[33] A plethora of other theories, including P-boxes,[34] lower previsions,[35] Dempster–Shafer

[27] Thomas Heath, 1921, *A History of Greek Mathematics*: vol. 1: *From Thales to Euclid*;
vol. 2: *From Aristarchus to Diophantus* Clarendon Press, Oxford, UK. Reissued by Dover
Publications, New York, 1981. Jacob Klein, 1992, *Greek Mathematical Thought and the
Origin of Algebra*, Dover Publications, New York.

[28] James Franklin, 2001, *The Science of Conjecture: Evidence and Probability before Pascal*,
The Johns Hopkins University Press, Baltimore.

[29] Ian Hacking, 1975, *The Emergence of Probability: A Philosophical Study of Early
Ideas About Probability, Induction and Statistical Inference*, Cambridge University Press,
Cambridge.

[30] Theodore M. Porter, *The Rise of Statistical Thinking: 1820–1900.* Princeton
University Press, Princeton, New Jersey. Stephen M. Stigler, 1986, *The History of
Statistics: The Measurement of Uncertainty before 1900*, Harvard University Press, Boston.

[31] C. Lejewski, 1996, "Jan Lukasiewicz," in P. Edwards, ed., *The Encyclopedia of
Philosophy*, vol. 5, Simon and Schuster Macmillan, New York.

[32] L. A. Zadeh 1965, "Fuzzy sets", *Information and Control*, 8: 338–53.

[33] A. Wald, 1945, "Statistical decision functions which minimize the maximum risk", *Annals of Mathematics*, 46(2): 265–80.

[34] S. Ferson, 2002, *RAMAS Risk Calc 4.0 Software: Risk Assessment with Uncertain
Numbers*, Lewis Publishers, Boca Raton.

[35] P. Walley, 1996, "Measures of uncertainty in expert systems", *Artificial Intelligence*, 83: 1–58. E. Miranda, 2008, "A survey of the theory of coherent lower
previsions", *Intl. J. Approximate Reasoning*, 48: 628–58.

theory,[36] generalized information theory,[37] and info-gap theory,[38] have continued to sprout up.

This suggests that theories of uncertainty will continue to grow and diversify. How many uncertainty theories do we need? Lots, and forever. Would we say that of physics? No, at least not forever.

Uncertainty thrives in the realm of imagination, incongruity, and contradiction. Uncertainty falls in the realm of science fiction as much as in the realm of science. Uncertainty accompanies assertions that you do not believe to be true, like Alice's skepticism about the Queen's assertion of her age (p. 26). Theories of uncertainty must be able to model and manage inconsistent, incoherent, or erroneous ideas.

To get a better appreciation of the challenge facing theories of uncertainty, let's consider some incongruous thoughts that a theory of uncertainty might be expected to accommodate. We'll begin with Humpty Dumpty's explanation, as reported by Lewis Carroll, of words that are "like a portmanteau—there are two meanings packed up into one word." For example, "'slithy' means 'lithe and slimy'" or "'mimsy' is 'flimsy and miserable'".[39]

Now let those two meanings be contradictory. A prosaic example might be: thinking the unthinkable, whose portmanteau word might be "unthinkinkable". We can hold in our minds thoughts of four-sided triangles (a logical contradiction), parallel lines that intersect (impossible in Euclidean geometry, but not in Lobachevskian geometry), and endless other seeming

[36] A. P. Dempster, 1967, "Upper and lower probabilities induced by multivalue mappings", *Annals of Mathematical Statistics*, 38: 325–39. G. Shafer, 1976, *A Mathematical Theory of Evidence*, Princeton University Press, Princeton, New Jersey.

[37] G. J. Klir, 2006, *Uncertainty and Information: Foundations of Generalized Information Theory*, John Wiley & Sons, Hoboken, New Jersey.

[38] Yakov Ben-Haim, 2006, *Info-gap Decision Theory: Decisions Under Severe Uncertainty*, 2nd edn., Academic Press, London.

[39] Lewis Carroll, *Through the Looking-Glass*, chapter 6.

impossibilities from super-girls like Pippi Longstocking[40] to life on Mars (some of which may actually be true, or at least physically possible).

Scientists, logicians, and saints are in the business of dispelling all such incongruities, errors, and contradictions. Banishing inconsistency is possible in science because (or if) there is only one world and it is logically coherent. Belief in one coherent world and one grand unified theory is the modern secular version of the ancient monotheistic intuition of one universal God (in which saints tend to believe). Uncertainty thrives in the realm in which scientists and saints have not yet completed their tasks (perhaps because they cannot be completed). For instance, we must entertain a wide range of conflicting conceptions when we do not yet know how (or whether) quantum mechanics can be reconciled with general relativity, or Pippi's strength reconciled with the limitations of physiology. As Henry Adams wrote:

> Images are not arguments, rarely even lead to proof, but the mind craves them, and, of late more than ever, the keenest experimenters find twenty images better than one, especially if contradictory; since the human mind has already learned to deal in contradictions.[41]

The very idea of a rigorously logical theory of uncertainty is startling and implausible because the realm of the uncertain is inherently incoherent and contradictory. The first uncertainty theory—probability—emerged many centuries after the invention of the axiomatic method in mathematics. Today we have many theories of uncertainty, as mentioned earlier. Why such a long and diverse list? It seems that in constructing a logically consistent theory of the logically *in*consistent domain of uncertainty,

[40] Astrid Lindgren, 1950, *Pippi Longstocking*, Viking Press, New York.

[41] Henry Adams, 1918, *The Education of Henry Adams*, edited by Ernest Samuels, Houghton Mifflin Co., Boston, p.489.

one cannot capture the whole beast all at once (though I'm uncertain about this).

The realm of uncertainty contains contradictions (ostensible or real) such as the pair of statements: "Nine-year-old girls can lift horses" and "Muscle fiber generates tension through the action of actin and myosin cross-bridge cycling". A logically consistent theory of uncertainty can handle improbabilities, as can scientific theories like quantum mechanics. But a logical theory cannot encompass outright contradictions. Science investigates a domain: the natural and physical worlds. Those worlds, by virtue of their existence, are perhaps coherent in a way that can be reflected in a unified logical theory. Theories of uncertainty are directed at a larger domain: the natural and physical worlds and all imaginable (and unimaginable) other worlds. That larger domain is definitely *not* logically coherent, and a unified logical theory would seem to be unattainable. Hence many theories of uncertainty would seem to be needed.

Scientific theories are good to have, and we do well to encourage the scientists. But it is a mistake to think that the scientific paradigm—with its insistence on logical consistency—is suitable to all domains, particularly to the study of uncertainty. Logic is a powerful tool and the axiomatic method assures the logical consistency of a theory. For instance, Leonard Savage argued that personal probability is a "code of consistency"[42] for choosing one's behavior. However, Jim March compares the rigorous logic of mathematical theories of decision to strict religious morality. Consistency between values and actions is commendable, says March, but he notes that one sometimes needs to deviate from perfect morality. While "[s]tandard notions of intelligent choice are theories of strict morality ... saints are a luxury to be

---

[42] Leonard J. Savage, 1972, *The Foundations of Statistics*, 2nd revised edn., Dover Publications, New York, p. 59.

encouraged only in small numbers."[43] Logical consistency is a merit of any single theory, including a theory of uncertainty. However, insisting that the same logical consistency apply over the entire domain of uncertainty and its diverse theories is like asking reality and saintliness to make peace.

The End of Science?

Innovation dilemmas result from uncertainty, and uncertainty is discordant, incongruous, and rowdy. Alice doubted the Queen's age (p. 26), but can one really doubt the endlessness of the search for scientific knowledge? Isn't science a perpetual search for truth? The answer will deepen our appreciation of uncertainty.

Scientific theories are good to have because they tend to reduce our uncertainty and augment coherence in our understanding of the natural and physical world. Science is the search for and study of patterns and laws in the world. Could that search become exhausted, like an overworked coal vein, leaving nothing more to be found? Could science end? After briefly touching on several fairly obvious possible endgames for science, we explore how the vast unknown could undermine—rather than underlie—the scientific enterprise. But ironically, we will conclude that the possibility that science could end is linked to the reason that science is possible at all. Our goal is to gain further insight into the idea of boundless uncertainty.

We do not know what surprises remain hidden among nature's secrets: an unknown galaxy, or a new subatomic force, or something about the essence of time and space. Science is the process of discovering these unknowns. Perhaps the most elusive

---

[43] James G. March, 1988, "Bounded rationality, ambiguity, and the engineering of choice". In: David E. Bell, Howard Raiffa, and Amos Tversky, eds., *Decision Making: Descriptive, Normative, and Prescriptive Interactions*, Cambridge University Press, Cambridge, p. 51.

unknown is the extent of nature's secrets. It is possible that the inventory of nature's unknowns is finite or conceivably even nearly empty. However, a look at open problems in science, from astronomy to zoology, suggests that nature's storehouse of surprises is still chock full. So, from this perspective, the answer to the question "Could science end?" is conceivably "Yes", but most probably "No".

Another possible "Yes" answer is that science will end by reaching the limit of human cognitive capability. Nature's storehouse of surprises may never empty out, but the rate of our discoveries may gradually fall, reaching zero when scientists have figured out everything that humans are able to understand. This is possible, but judging from the last 400 years, it seems that we've only just begun to tap our mind's expansive capability.

Or perhaps science—a product of human civilization—will end due to historical or social forces. The simplest such scenario is that we blow ourselves to smithereens. Smithereens can't do science. Another more complicated scenario is Oswald Spengler's theory of cyclical history,[44] whereby an advanced society—such as Western civilization—decays and disappears, science disappearing with it. So again a tentative "Yes". But this might only be an interruption of science if later civilizations resume the search.

We now explore the main mechanism by which science could become impossible. This will lead to deeper understanding of the relation between knowledge and the unknown and to understanding why science is possible at all.

One axiom of science is that there exist stable and discoverable laws of nature. As the philosopher A. N. Whitehead wrote: "Apart from recurrence, knowledge would be impossible; for nothing could be referred to our past experience. Also, apart from some

---

[44] Oswald Spengler, *Decline of the West*, vol. 1: *Form and Actuality*, 1926, vol. 2: *Perspectives of World History*, 1928, Alfred A. Knopf, New York.

regularity of recurrence, measurement would be impossible."[45] The stability of phenomena is what allows a scientist to find patterns in nature and to repeat, study, and build upon the work of other scientists. Without regular recurrence there would be no such thing as a discoverable law of nature.

However, as David Hume explained long ago,[46] one can never empirically prove that regular recurrence will hold in the future. If one tries to test the regularity of the future, one finds that the future has become the past. The future can never be tested, just as one can never step on the rolled-up part of an endless rug unfurling always in front of you.

Suppose the axiom of natural law turns out to be wrong, or suppose nature comes unstuck and its laws start "sliding around", changing. Science would end. If regularity, patterns, and laws no longer exist, then scientific pursuit of them becomes fruitless.

Or maybe not. Couldn't scientists search for the laws by which nature "slides around"? Quantum mechanics seems to do just that. For instance, when a polarized photon impinges on a polarizing crystal, the photon will either be entirely absorbed or entirely transmitted, as Dirac explained.[47] An individual photon's fate is not determined by any law of nature (if you believe quantum mechanics). Nature is indeterminate regarding an individual photon in this situation. Nonetheless, quantum theory very accurately predicts the probability that the photon will be transmitted, and the probability that it will be absorbed. In other words, quantum mechanics establishes a deterministic law (a known and fixed probability distribution) describing nature's indeterminism about the fate of individual photons.

[45] Alfred North Whitehead, 1925, *Science and the Modern World*, Lowell Lectures. 1948, Mentor Books, New York, p. 36.

[46] Hume, 1748, *An Enquiry Concerning Human Understanding*.

[47] P. A. M. Dirac, 1958, *The Principles of Quantum Mechanics*, 4th edn., Oxford University Press, London, pp. 4–7.

Suppose nature's indeterminism itself becomes lawless. Is that conceivable? Could nature become so disorderly, so confused and uncertain, so "out of joint: O, cursed spite", that no law can "set it right"?[48] The answer is conceivably "Yes", and if this happens then scientists are all out of a job. To understand how this is conceivable, one must appreciate the unknown at its most rambunctious.

Let's take stock. We can identify attributes of nature that are necessary for science to be possible. The axiom of natural law is one necessary attribute. The successful history of science suggests that the axiom of natural law has held firm in the past. But Hume assures us that we cannot know for sure that it will do so in the future.

In order to understand how natural law could come unstuck, we need to understand how natural law works (today). When a projectile, say a baseball, is thrown from here to there, its progress at each point along its trajectory is described, scientifically, in terms of its current position, direction of motion, and attributes such as its shape, mass, and surrounding medium. The laws of nature enable the calculation of the ball's progress by solving a mathematical equation whose starting point is the current state of the ball.

We can roughly describe most laws of nature as formulations of problems—for example as mathematical equations—whose input is the current and past states of the system in question, and whose solution predicts an outcome: the next state of the system. What is law-like about this is that these problems—whose solution describes a progression, like the flight of a baseball—are constant over time. The scientist calculates the baseball's trajectory by solving the same problem over and over again (or all at once with a differential equation). Sometimes such problem are

48 William Shakespeare, *Hamlet*, Act I, scene v, line 190.

hard to solve, so scientists are good mathematicians or they have big computers (or both). But solvable they are.

Let's remember that nature is not a scientist, and nature does not solve a problem when things happen (like baseballs speeding toward home plate). Nature just *does it*. The scientist's law is a description of nature, not nature itself.

There are other laws of nature for which we must modify the previous description. In these cases the law of nature is, as before, the formulation of a problem. Now, however, the solution of the problem not only predicts the next state of the system, but it also re-formulates the problem that must be solved at the next step. There is a sort of feedback: the next state of the system alters some part of the rule by which subsequent progress is made. For instance, when an object falls toward earth from outer space, the rate at which the object falls depends on the strength of the gravitational attraction. The strength of the gravitational attraction, in turn, increases as the object gets closer. Thus the details of the problem to be solved change as the object moves. Problems like these tend to be more difficult to solve, but that's the scientist's problem (or pleasure). There is of course something stable about the gravitational interaction, but some details—its specific magnitude—change during the fall.

Now we can appreciate how nature might become lawlessly unstuck. Let's consider the second type of natural law, where the problem—the law itself—gets modified by the evolving event. Let's furthermore suppose that the problem is not simply difficult to solve, but that no solution can be obtained in a finite amount of time (computer scientists have lots of examples of uncomputable problems like this). As before, nature itself does not solve a problem; nature just *does it*. But the scientist is now in the position that no prediction can be made, no trajectory can be calculated, no model or description of the phenomenon can be obtained—no explicit solution of the problem statement exists.

This is because the problem to be solved evolves continuously from previous solutions, and none of the sequence of problems can be solved. The scientist's profession will become frustrating, futile, and fruitless.

Nature becomes lawlessly unstuck, and science ends, if all laws of nature become of the modified second type. The world itself will continue because nature solves no problems, it just does its thing. But the *way* it does this is now so raw and unruly that no study of nature can get to first base.

Does this sound like science fiction (or nightmare)? Maybe. But as far as we know, the only thing between us and this new state of affairs is an unprovable prior assumption: the axiom of natural law. Scientists assume that laws exist and are stable because past experience, together with our psychological makeup, very strongly suggest that regular recurrence can be relied upon. But if you think that the scientists can empirically *prove* that the future will continue to be lawful, like the past, recall that *all* experience is *past* experience. Recall the unfurling-rug metaphor, and make an appointment to see Mr Hume.

Is science likely to become fruitless or boring? No. Science is possible because there is (or seems to be) a real world "out there" with stable universal patterns. However, science is also possible because there are things we don't know but can discover. Science thrives on an unknown that is full of surprises, one of which might be that there is nothing more to discover. Science—the search for natural laws—thrives even though the future existence of natural law can never be proven. Science thrives for the very same reason that we can never be sure that science will not someday end. Science thrives because of uncertainty, which also underlies the innovation dilemmas that confront us on all sides.

I am a Believer

With some practice, the White Queen was able to believe six impossible things before breakfast (p. 26), so she would have easily understood the secrets of uncertainty, ignorance, and surprise. She would also have had no difficulty in recognizing innovation dilemmas because they result from the vastness of the unknown. Nonetheless, some things I believe without reservation.

Science has made great progress, but there are many things that I don't know. *About the past*: how my great-great-grandfather supported his family, how Charlemagne consolidated his imperial power, or how Rabbi Akiva became a scholar. *About the future*: whether I'll get that contract, how much the climate will change in the next hundred years, or when the next war will erupt. *About why things are as they are*: why stones fall and water freezes, or why people love or hate or don't give a damn, or why we are, period.

We reflect about questions like these, trying to answer them and to learn from them. For instance, we are interested in the relations between Charlemagne and his co-ruling brother Carloman. This can tell us about brothers, about emperors, and about power. We are interested in Akiva because he supposedly learned to read and started studying Torah at the age of forty, which tells us something about the indomitable human spirit.

We sometimes get to the bottom of things and understand the whys and ways of our world. We see patterns and discover laws of nature, or at least we tell stories about how things happen. Stones fall because it's their nature to seek the center of the world (Aristotle), or due to gravitational attraction (Newton), or because of mass-induced space warp (Einstein). Human history has its patterns, driven by the will to power of heroic leaders, or by the unfolding of truth and justice, or by God's hand in history.

We also think about thinking itself, as suggested by Rodin's statue of *The Thinker*. What *is* thinking (or what do we think it is)? Is thinking a physical process with electrons whirling in our brain? Or does thinking involve something transcendental; maybe the soul whirling in the spheres? Each age has its answers.

We sometimes get stuck, and can't figure out or get to the bottom of things. Sometimes we even realize that there is no "bottom"; that each answer brings its own questions. As John Wheeler said, "We live on an island of knowledge surrounded by a sea of ignorance. As our island of knowledge grows, so does the shore of our ignorance."[49]

Sometimes we get stuck in an even subtler way that is very puzzling, and even disturbing. Any rational chain of thought must have a starting point. Any rational justification of that starting point must have its own starting point. In other words, any attempt to rationally justify rational thought can never be completed. Rational thought cannot justify itself, which is *almost* the same as saying that rational thought is not justified. Any specific rational argument—Einstein's cosmology or Piaget's psychology—is justified based on its premises (and evidence, and many other things). But Rational Thought, as a method, as a way of life and a core of civilization, cannot ultimately and unequivocally justify itself.

I believe that experience reflects reality, and that thought organizes experience to reveal the patterns of reality. The truth of this belief is, I believe, self evident and unavoidable. Just look around you. Flowers bloom anew each year. Planets swoop around with great regularity. We have learned enough about our small planetary corner of the universe to change it, to control it, to benefit from it, even to greatly endanger it. I believe that rational thought is justified, but that's a belief, not a rational argument.

[49] John A. Wheeler, quoted in *Scientific American*, December 1992, p. 20.

Rational thought, in its many different forms, is not only justified; it is unavoidable. We can't resist it. Moses saw the flaming bush and was both frightened and curious because it was not consumed.[50] He was drawn to it despite his fear. The unknown draws us irresistibly on an endless search for order and understanding. The unknown drives us to search for knowledge, and that search is not fruitless. This I believe.

[50] Exodus 3:1–3.

4

Optimization and Its Limits

"'What day of the month is it?' he said, turning to Alice: he had taken his watch out of his pocket, and was looking at it uneasily, shaking it every now and then, and holding it to his ear.

"Alice considered a little, and then said 'The fourth.'

"'Two days wrong!' sighed the Hatter. 'I told you butter wouldn't suit the works!' he added looking angrily at the March Hare.

"'It was the *best* butter,' the March Hare meekly replied."[1]

We have completed our discussion of uncertainty arising from the boundless potential for discovery, invention, and surprise. The second element that often impacts an innovation dilemma is the deep-rooted tendency to seek optimal—that is, minimal or maximal—outcomes. We will see that uncertainty sometimes makes outcome optimization infeasible or even unwise, and leads us to seek a different approach.

The wonders of human progress owe their origin to our thirst for improvement. However, our ability to predict the outcome quality of the options among which we must choose is limited when our knowledge and understanding are severely curtailed. We simply can't know which option will be optimal. In addition,

---

[1] Lewis Carroll, *Alice's Adventures in Wonderland*, chapter 7.

The Dilemmas of Wonderland: Decisions in the Age of Innovation. Yakov Ben-Haim.
© Yakov Ben-Haim 2018. Published in 2018 by Oxford University Press.
DOI: 10.1093/oso/9780198822233.001.0001

seeking the best outcome sometimes becomes a moral imperative of its own, regardless of substantive needs and values. Optimization is then a goal in itself, and this may lead to misuse of an otherwise worthy idea.

We begin with a discussion of the modern paradigm of optimization—the laws of physics—and then proceed to discuss three ways in which the search for optimal solutions may go astray because optimization has become an imperative of its own. After discussing the distinction between substantive and procedural optimization, we will see that the outcome optimizer inadvertently minimizes the robustness against uncertainty.

Physics: The Paradigm of the Optimum

A hanging bridge, suspended across a gorge, takes a particular shape, and not any other. Does a principle determine that shape, and does the same principle govern a clothes line loaded with wet towels, and the majestically sloped branches of a willow tree? Galileo's stone gathered speed at a fixed rate after he dropped it from the tower of Pisa; a pebble tossed by a child follows a special trajectory; a swallow with finely shaped wings swoops along a particular path: does one universal principle govern all physical dynamics? The refraction of light in a prism; the quantum tunneling of radioactive alpha decay; the spinning expansion of the cosmos: does one principle generate the laws governing all of these phenomena?

Yes, and that principle is an optimization. We'll leave it to the mathematical physicists to work out the details, but a simple example will explain the general idea.

Imagine the shape of a hanging bridge, suspended from the walls of a gorge. The gravitational attraction between the earth and the bridge pulls the bridge downward as far as possible, limited by the length of the walkway and the width of the gorge.

We already see an optimization emerging in the phrase "as far as possible". The downward displacement of the bridge is not maximal at each point, but it is maximal over the bridge as a whole, which we can understand as follows.

If the bridge were disconnected from the gorge, it would fall and hit the ground with some amount of energy. That is, the hanging bridge has the potential to gain energy. In its hanging state it has "potential energy". Now here comes the optimization. Gravity pulls the bridge downward as far as possible, giving it a shape whose potential energy is minimal. Any other shape—a kink here or an undulation there—would have greater potential energy. Equivalently, any other shape would enable the disconnected bridge to achieve greater kinetic energy as it falls to the ground. The earth interacts with the hanging bridge, through their mutual gravitational attraction, to achieve a shape with lower potential energy than any other shape. The shape is optimal in the sense that it minimizes the potential energy.

The same principle of optimization applies to the clothes line and the willow tree and to all other static gravitational interactions. The example of minimum potential energy is a special case of a much more general principle, the law of least action, that governs the entire domain of physics. The specific equations that govern the shape of hanging bridges, or the equations that describe the flight of a photon or the swirl of a galaxy, are all derivable by finding the minimum of what the physicists call the action integral.[2] The laws of physics are obtained by solving optimization problems.

One of the distinctive and important features of optimization problems is that, when they are solvable, their solutions

[2] Cornelius Lanczos, 1970, *The Variational Principles of Mechanics*, 4th edition, Dover Publications, New York.

are unique.[3] After specifying the details of the problem, a single solution determines a unique behavior of the system. The bridge hangs in one and only one shape; the pebble soars through the air on a unique trajectory; the quantum-mechanical wave function describes a single time–space evolution of a quantum system,[4] and so on. The solution is unique and optimal, where these two attributes are intimately linked: it is the "best" solution (minimal action), better than all others, and hence unique.

A solution that is better than all other solutions is, by definition, unique: it is the one and only best solution. The interesting thing is that this can be reversed: something that is unique must be an optimum according to some criterion. Something that is unique has some attribute (e.g. potential energy for hanging bridge shapes) that is different from all other conceivable solutions. The distinct attribute of a unique solution can always be characterized, in one way or another, as an optimum. Any situation that must provide a unique solution must do so by being the solution of an optimization.

The uniqueness of optimal solutions removes the need for value judgments in choosing a solution. Once the criterion for optimization is specified, one does not need to make any judgment of how good the solution is, or how much better it is than another. The criterion of the optimization determines the solution uniquely. Natural law, based on optimization, needn't

[3] We will ignore the technical complication of multiple optima: solutions that are all equally good. Such complications are obviated by seeking a set of optimal solutions, and such a set *is* unique.

[4] One might think that quantum mechanics provides counterexamples, but the physicists have shown that quantum mechanical wave functions satisfy the least-action principle, just like all other physical systems. See Richard P. Feynman, 1948, "Space-time approach to non-relativistic quantum mechanics," *Reviews of Modern Physics*, 20(2), pp. 367–87.

assume purposefulness in nature, while still assuring the continual progression from one event to the next.

In contrast, Aristotle wrote that "Nature, as we often say, makes nothing in vain" and that its patterns are teleological, directed to a goal: "the final cause and end of a thing is the best".[5] For Leibniz, this is the best of all possible worlds that God could have created.[6] For modern physics, however, nature "works" by solving an optimization problem and nature is "best" only in the value-neutral sense of minimizing the action integral.

The ability of optimization to provide unique solutions without the need for subjective value judgments has made it attractive in economics and social science. The "rational actor" model assumes that economic or political decision makers choose their actions to maximize their utility (or their average utility). For example, Krepinevich and Watts write:

> Many RAND analysts [in the 1950s and '60s] assumed that the Soviets were supremely rational planners—that their major decisions about military forces were carefully calculated to pose the greatest threat to the United States in general, and to SAC [Strategic Air Command] in particular. This construct had great appeal at RAND and elsewhere in the US national security establishment because it allowed analysts to simplify their assumptions about Soviet behavior. By assuming the Soviets would behave rationally (at least as viewed from a US perspective), one could forecast future Soviet strategic forces without having to delve into the history, propensities, strengths and weaknesses, rigidities, military doctrines, operational methods, and organizational complexities of the Soviet state.[7]

[5] Aristotle, *Politics*, $1253^a 1$, 8. See also $1256^b 20$.

[6] Bertrand Russell, 1937, *A Critical Exposition of the Philosophy of Leibniz.* George Allen and Unwin, London, 2nd edn. pp. 36–7.

[7] Andrew F. Krepinevich and Barry D. Watts, 2015, *The Last Warrior: Andrew Marshall and the Shaping of Modern American Defense Strategy.* Basic Books, New York, p. 44.

Some optima can be calculated or known in other ways before they are observed. Physics provides many examples by minimizing the action integral as we explained earlier. However, some optima are unknowable because of Shackle–Popper indeterminism. We may know or believe that an optimum exists, but we can never know its value.

Consider, for example, the 100-meter sprint. The men's world record, recognized by the International Association of Athletics Federations as of August 2016, is 9.58 seconds, set by Jamaica's Usain Bolt in 2009. The previous record was 9.69 seconds, set by Bolt in August 2008, and before that the record was 9.72 seconds, set by Bolt in May 2008. Before that, Asafa Powell set the record of 9.74 seconds in September 2007.

What is the minimal time required to run 100 meters? It's certainly not zero seconds, nor would one believe that, say, five seconds would suffice. But is the optimal time 8.53 seconds, or 9.27 seconds? We really cannot know, in part because we cannot know what innovations of health, hygiene, training, nutrition, or running-shoe design will appear, or what other imaginable and as yet unimagined innovations will occur that can further shorten the run time. Shackle–Popper indeterminism precludes the possibility of knowing the best possible record. The optimum presumably exists, in the sense that we are confident that nobody will ever run 100 meters in less than five seconds. However, we cannot know what that optimum is, nor will we ever know it, even if some day a record holds for years and years without being overturned.

Similar arguments about unknowable optima apply to the maximum payload of nuclear weapons, the minimum size of house-cleaning robots, and endless other examples (many of which have yet to be invented). Optima surely exist, but we can never know their values. Shackle–Popper indeterminism prevents us from knowing the future innovations that will someday arise and how they will enhance the best currently available solutions.

Physics provides the modern paradigm of optimization, and we now begin to explore the nature of its use, and the extent of its suitability, in human affairs.

Optimization and Its Uses

The concept of optimization is widely used in three interrelated and overlapping forms: idealistic, rhetorical, and substantive.[8] The uncertainty that often accompanies our decisions, especially when facing an innovation dilemma, implies that these uses are not always feasible, reliable, or wise.

The **idealism** of optimization uses the language of maximizing or minimizing to express the laudable desire to achieve the best possible outcome. Jeremy Bentham wrote the "pleasing truth" that:

> The age we live in is a busy age; in which knowledge is rapidly advancing towards perfection. In the natural world, in particular, every thing teems with discovery and with improvement.

Bentham continues that, "Correspondent to *discovery* and *improvement* in the natural world, is *reformation* in the moral", and he reveals the "fundamental axiom" for this reformation: *"it is the greatest happiness of the greatest number that is the measure of right and wrong".*[9] Bentham is expressing a social or civilizational attitude that things can be better, even approaching perfection, and it is our responsibility to make them so to the fullest extent possible.

---

[8] We will consider a fourth use of the concept, procedural optimization, on p. 76.

[9] Jeremy Bentham, 1776, *A Fragment on Government*, appearing in Ross Harrison, ed., *Selected Writings on Utilitarianism: Bentham*, Wordsworth Classics, Hertfordshire, UK, 2000, p. 3. Emphasis in the original.

While not necessarily condemning the past for its failures, the optimization idealist may look condescendingly or uncomprehendingly on the past. Slavery in Periclean Athens, for example, may be seen as a primitive hold over, inconsistent with the democratic principles that were emerging. Twenty-first-century Westerners may sympathize with their earlier compatriots upon learning that the average life expectancy in the US was less than fifty years in 1900; nobody's fault, but we've gotten much better at health and hygiene since then.

The idealism of optimization is illustrated in the psychiatric report on the bipolar woman who wished to bear a healthy child, discussed in Chapter 2. The psychiatrists sought "to *maximize the likelihood* that maternal mood and behavioral stability will be maintained". The patient "wanted to become pregnant with the *safest medications possible.*" Consequently, "the decision was made to use olanzapine to *maximize mood stability*". The therapists and the patient recognized the need for "potentially teratogenic medications that may result in adverse side effects for both mother and fetus."[10] Nonetheless, the severe uncertainties did not deter the psychiatric team or the patient from expressing the idealistic aspiration for optimal outcomes for both mother and child.

The **rhetoric** of optimization takes various forms, where the common denominator is ambiguous reference to optimality that obscures far-reaching implications. To say that "This is the best solution" is a commendation, even though this best solution may nonetheless be terrible. The phrase "as good as it gets" sometimes obscures the reality of a pretty bad situation that won't get any better. The movie *As Good As It Gets*, with Jack Nicholson and Helen Hunt, is the romantic story of two people who develop a relationship that is an improvement for both of them, but that is far from perfect and won't get much better.

---

[10] Burt *et al.*, 2010, *op. cit.*, pp. 892–4. Emphasis added.

The rhetoric of optimization sometimes attempts to strengthen an otherwise weak or incomplete argument, or to cover up imprecise reasoning by association with the venerated intellectual tradition of optimization. For example, financial investment firms commonly promise to maximize your returns. *Forbes* discusses "5 Best Ways to Maximize Your Retirement Investments"[11] (note the double optimization). The article itself focuses on the more modest goal of having "enough money in retirement", but the rhetoric of optimization has caught attention and covered the gap between aspirations and reality.

Rhetoric and expressions of idealism sometimes get entangled. Bentham's famous axiom of morality that we discussed earlier— "the greatest happiness of the greatest number"—is an example. The dictum expresses an ideal, an aspiration, and Bentham devoted great effort in transforming this ideal into the coherent logical philosophy that he called Utilitarianism. The axiom itself, however, is ambiguous. The double optimization, greatest happiness and greatest number, engenders an infinity of solutions: a few very happy people or many moderately happy people can have the same overall happiness. The rhetoric of double optimization appears to reconcile a conflict that is inherent in allocating a finite amount of happiness, while no such reconciliation is at all self evident. John Stuart Mill's defense of Utilitarianism, nearly a century after Bentham, recognized this difficulty and proposed a solution:

> As the means of making the nearest approach to this ideal, utility [Bentham's theory of Utilitarianism] would enjoin, first, that laws and social arrangements should place the…interest, of every individual, as nearly as possible in harmony with the

[11] *Forbes* online, March 16, 2012, accessed 22.10.2015: http://www.forbes.com/sites/thestreet/2012/03/16/5-best-ways-to-maximize-your-retirement-investments.

interest of the whole; and secondly, that education and
opinion...should...establish in the mind of every individual
an indissoluble association between his own happiness and the
good of the whole.[12]

This aspiration for uniformity of interests, reflecting the socialist
side of Mill's thought, tends to undermine individual freedom and
social diversity. This example illustrates the rhetoric of optimization: Bentham's expression of double optimality was ambiguous
and obscured far-reaching social and political implications that
may be undesirable.

Substantive optimization is a procedure that uses assumptions, knowledge, data, rules of reasoning, and so on, to select
a solution or action whose predicted outcome is best according
to a specified criterion of what is good or valuable. The emphasis
of substantive optimization is on the value of the outcome, and
is distinguished from procedural optimization that we discuss
shortly.

Quantitative procedures for substantive optimization entail
explicit mathematical algorithms for identifying optimal solutions, such as differential calculus, Bellman's method of dynamic
programming,[13] or tools based on the algebra of vectors.[14]
Substantive optimization in the qualitative realm is also well
developed, though in very different ways. For instance, Thomas
Fingar, former Deputy Director of National Intelligence in the
US intelligence community, advocates identifying "best-case
and worst-case judgments that frame estimates of more likely

[12] John Stuart Mill, 1861, *Utilitarianism.* Appearing in Edwin A. Burtt, ed., 1939,
The English Philosophers From Bacon to Mill, The Modern Library, New York, p. 908.

[13] Richard Bellman, 1961, *Adaptive Control Processes: A Guided Tour*, Princeton
University Press, Princeton, NJ.

[14] David G. Luenberger, 1969, *Optimization by Vector Space Methods*, Wiley, New
York.

possibilities."[15] At a more fundamental level, Plato has Socrates arguing that systematic rational thought protects the mind from the illusions of the senses.[16] Similarly, Aristotle claimed that the nature of things can be revealed only by speech that is subject to the discipline of logic.[17]

Substantive optimization sometimes ignores important aspects of the problem in formulating or justifying the optimization. Engineers tend to optimize a specific design concept without looking at the complete realm of design possibilities. For instance, pressurized-water nuclear reactors are optimized within the realm of possibilities that precludes boiling of the coolant. In contrast, boiling-water reactors are optimized within a design space that presumes coolant boiling. Both of these classes of reactors are perhaps optimal—each in its own conceptual domain—in their exploitation of the properties of enriched uranium fuel. In contrast, the CANDU reactor concept is optimized for *un*enriched uranium. Optimization in light of a "big picture" is not pursued. Each of these three designs is a "local" optimum: better than similar alternatives in the relevant design space. One of them (or perhaps none) may be a "global" optimum: better than all alternatives. Engineering designers are well aware of the local/global optimum dichotomy, but it's a major challenge to implement global optimization in practice.[18] Furthermore, the

---

[15] Thomas Fingar, 2011, *Reducing Uncertainty: Intelligence Analysis and National Security*, Stanford Security Studies, Stanford, CA.

[16] Plato's *Republic*, 10.602e–603b.

[17] Jacob Klein, "Aristotle: an Introduction", pp. 171–95 in Robert B. Williamson and Elliot Zuckerman, eds., 2013, *Jacob Klein Lectures and Essays*, St John's College Bookstore, St John's College Press, Winnipeg, Manitoba, Canada. See p. 178.

[18] Part of the problem is defining the global realm of possibilities. Does one optimize over these three concepts (PWR, BWR, CANDU), or over all possible nuclear reactor designs, or over all possible fuel-based designs, or all possible energy-generating systems? Does one consider externalities such as social or

rhetoric of optimization may elide the local/global distinction when the final product reaches the sales and promotion stage.

Substantive optimization is sometimes implemented by presuming the possession of critical knowledge that we don't actually have. Economics provides an example. Analysts of modern economies are well aware of the frequency of surprises resulting from unique, unforeseen, and unprecedented political, social, or technological events or innovations. The peaceful dissolution of the Soviet Union in 1991, the "Arab Spring" in 2011, and the invention of portable computers are typical examples. Many quantitative economic models—intended for outcome-optimal policy formulation—assume that we know the relative likelihoods of these unanticipated events or innovations, and this in turn presumes that we know what those events or innovations could be. One cannot know something that is unforeseen and unprecedented. Consequently, policy formulation based on presumed (but in fact absent) knowledge may not satisfy the claim of optimality. This misuse of substantive optimization results from ignoring, perhaps innocently, Knightian uncertainty and Shackle–Popper indeterminism, discussed in chapter 3.

Another misuse of substantive optimization arises in attempts to formulate optimal algorithms for learning and discovery at the boundary of human knowledge. By "learning" we mean discovery of new knowledge, unknown previously to those engaged in planning the learning process. This is unlike learning a subject (the multiplication table or the history of France) from a teacher who has mastered it.

Imagine a team of spelunkers exploring a newly discovered cave. Their goal, let us suppose, is to learn (that is, to discover) the depth of the cave. The planner wishes to minimize (optimize) the

environmental impact? Does one include social or institutional change—such as legislation regulating energy use—as part of the design space?

time required to discover the cave depth and return to sunlight. The greater the rate of progress through the cave, the shorter the duration until the learning has occurred. Progress is faster if less food is carried (due to weight), but progress is slower if food runs low (due to hunger). If we knew the depth of the cave we could calculate the amount of food to minimize the time in and back. But we don't know the depth of the cave; the expedition is intended to discover that missing fact. Direct minimization of the transit time is not possible because we lack—and seek to learn—the very datum that is needed for the optimization.

The optimizer might reasonably suggest that we minimize the average (rather than the actual) time to discover the cave depth. To do this we must know the probability distribution of cave depths. Many caves have been explored, and our probability distribution can be based on this data. But how many caves in the world have *not* been explored? Is our newly discovered cave from the same class as previously explored caves whose probability distribution we can estimate? Or are there in fact many very different types of caves, hidden in the depths of the mountains and under the seas? Are we really confident that our new cave is like other known caves? Perhaps, but we know very little about the new cave, so to assume so would be to ignore an essential element of what learning is all about: extending our knowledge of the unknown.

This simplified hypothetical spelunking example illustrates a fundamental contradiction in attempting to formulate algorithms for optimal discovery of new knowledge. The goal is laudable: to optimize the use of resources in reducing our ignorance about a subject of importance. For example, governments, firms, and individuals expend time and money to learn about the economy in order to make responsible decisions. The aspiration is to allocate these resources optimally: minimal time, maximal return, minimal cost, maximal public benefit, and so on.[19] However, in order to choose the allocation that optimizes the learning, one must know some unknown facts or mechanisms of the economy that is being studied. If we know those facts or mechanisms, we have nothing to learn or discover; if we don't know them, we can't optimize the knowledge-acquisition process. The optimizer might suggest stochastic optimization (like in the spelunking case), but that requires knowledge of a probability distribution which is rarely known, especially in dynamic innovative economies. Or, the optimizer might suggest an optimally adaptive learning algorithm: adjust the learning algorithm as your learning progresses. Adaptivity (and its close cousin, flexibility) can be very useful, but the claim to have an *optimal* adaptive algorithm for learning is problematic. There is most likely the lurking assumption of knowing some part of what is being learned *before* we learn it. For example, the adaptive algorithm updates the learning process at some rate (once a day, once a week, etc). Optimizing the update rate depends on the rate of learning, which cannot be known before we learn the (currently unknown) terrain of our ignorance.

We have discussed three overlapping uses of the concept of optimization: idealistic, rhetorical, and substantive. The thirst for improvement underlies many creative accomplishments of the human mind in all areas of endeavor. Without this idealistic drive to improve the human condition, we would still be living in damp caves and digging roots with our fingers, or worse. The ideal of optimization is rightly honored in the pantheon of civilization-building concepts. However, this idealization of optimization creates the danger of rhetorical exploitation of the idea to subtly introduce ambiguity or imprecise reasoning.

[19] George W. Evans and Seppo Honkapohja, 2001, *Learning and Expectations in Macroeconomics*, Princeton University Press, Princeton, NJ.

Furthermore, methods for optimizing the substantive value of an outcome are sometimes used while simplifying the problem, or perhaps even oversimplifying it, in order to ameliorate the practical challenges of finding optimal solutions. In addition, substantive optimization is sometimes used even when deep uncertainty precludes, in principle, optimization of the desired outcome. The moral or practical imperative to do one's best may obscure the fact that trying to optimize an uncertain outcome may be infeasible, unrealistic, or even irresponsible. Finally, trying to optimize the outcome of our decisions faces an irrevocable trade-off, as we discuss later in this chapter (p. 76).

The Innovation Dilemma Revisited

We can now understand how the optimization imperative adversely impacts an innovation dilemma by favoring the putatively optimal outcome, even though consideration of uncertainty might justifiably lead to a different choice.

The innovation dilemma arises from the uncertainty surrounding a new and unfamiliar—but promising—opportunity. The innovation purports to be better than existing options, but its newness enhances the uncertainty about the true outcome. The innovation promises to be an improvement, but it may actually turn out much worse than anticipated. We cannot dispel this uncertainty given our current knowledge; that's what uncertainty about the unknown is all about.

An optimistic belief in progress and improvement is probably necessary for—or at least very conducive to—the occurrence of progress and improvement. Belief in progress is a welcome self-fulfilling prophecy. Prophets of regression and doom rarely induce innovation and progress.

Having said that, rhetorically enhanced idealistic enthusiasm for optimization can propel one to ignore the implications of uncertainty. The fact of newness is not evidence of improvement, and the putative appearance of progress is a claim, not (yet) a fact.

The uniqueness of the optimum discussed earlier (p. 62) acts in a special way to magnify the innovation dilemma, as is now explained.

An outcome-optimal solution is the option that is predicted, based on our best knowledge at the planning or decision-making stage, to be better than all other options. The optimum might be thought (wrongly as we will see) to be like a unique mountain peak towering above a sea of uncertainty whose waves threaten to sweep away the outcome we desire. We are encouraged to select the predicted optimum because it seems to be most immune to uncertainty, towering as it were above all other alternatives. The problem with this argument is that the prediction itself is highly uncertain because our current knowledge is most likely wrong in fundamental but unknown ways. The fallacy resides in conflating the predicted quality of the outcome with an assumed immunity against ignorance, surprise, and uncertainty. An option that is predicted to have the best outcome is by no means the option that is most resistant to error in the prediction. These two attributes—predicted outcome-optimality and robustness against uncertainty—are fundamentally different. There is no reason to believe that, because an option is predicted to be best, its performance will be anywhere close to the prediction if and when reality turns out radically different from current understanding. Believing that the predicted optimal outcome has maximal immunity against uncertainty is wishful thinking, like believing that tasty food is always healthy.

We can further understand this limitation of outcome optimization by noting that the predicted optimum is found by exploiting all aspects of our current knowledge in order to predict the unique best alternative. The predicted optimum is in fact the most vulnerable to error in current knowledge, not the least vulnerable as suggested by the mountain metaphor. This is because seeking the optimum requires the exhaustive use of our knowledge—the valid as well as the erroneous parts. For instance, if we seek the option that is most likely to succeed, then we must use the entire probability distribution of relevant events, including the farthest tails of the distribution. The far tails, however, are least well known and reflect the rarest events, many of which may never have occurred or been observed. Similarly, if we seek the investment whose revenue is greatest then we must establish that all alternatives, even the most exotic and poorly known, are strictly worse under all possible future economic developments. Predicting the putative best outcome requires exploiting every nook and cranny of available knowledge, and this exposes that prediction to error on many fronts. When facing severe uncertainty, we should use our putative knowledge as sparingly as possible—not exhaustively—because important facets of that knowledge are wrong, but we don't yet know which, or by how much.

The metaphorical mountain of outcomes rising from the sea of uncertainty is an erroneously mixed metaphor. A better metaphor would imagine a sea of alternatives whose predicted outcomes are acceptable but whose real outcomes could vary greatly as our understanding changes. Out of this sea of uncertain alternatives rises a mountain of acceptable options whose behaviors are increasingly resistant to error in our knowledge. From among all options that are predicted (with current understanding) to be acceptable, we seek the unique alternative that is highest on this mountain of robustness against our ignorance. We are looking for an answer to the "robustness question". Namely, in considering any specific option, we ask: how wrong can we be in our current understanding, while ensuring that the outcome of that option will be acceptable? We are looking for the option with the greatest robustness against uncertainty.

This unique alternative is indeed an optimum, but it is not the optimum of a substantive good. It is not the option that purports to have the substantively best outcome. Rather, it is a **procedural** optimum: the unique solution that optimizes the robustness against surprise while achieving goals that are substantively adequate.[20] The robustness is not itself of substantive value once the outcome is achieved. Robustness is not a property of the outcome, but rather of the procedure. This is unlike the substantive value of an outcome that inheres to the outcome itself, like utility or cost. The outcome, once it occurs, does not have a property of robustness. Robustness inheres to the procedure of achieving a substantively valuable (though not necessarily optimal) outcome.

The imperative to strive for the best substantive outcome blurs the distinction between substantive and procedural optimization.[21] An innovation dilemma occurs when uncertainty is rampant. In this case, one should not automatically choose the putatively best outcome (because our predictions are highly unreliable) but rather the option that achieves an acceptable outcome with greatest robustness against uncertainty and surprise. Our understanding of this conclusion is deepened by appreciating an irrevocable trade-off that will now be explained.

Trade-Off: Confidence or Quality

We admire excellence in all areas of endeavor: art, sport, science, business, and daily life. The fastest runner wins the race and

[20] The concepts of substantive and procedural optimization are related to Simon's concepts of substantive and procedural rationality. See Herbert Alexander Simon, 1997, *Models of Bounded Rationality*, Vol. 3, *Empirically Grounded Economic Reason*, MIT Press, Cambridge, MA, p. 369.

[21] We discussed the idea of substantive optimization on p. 68.

our admiration. The lowest-cost design—all else being equal—
is preferred. "Better", by definition, is "more desirable" and—by
the logic of preference—the best is most preferred. The logic of
preference is so compelling that there sometimes seems to be
a moral imperative to do our best. However, "doing our best"
must be understood in light of an irrevocable trade-off between
the quality of the outcome that we aspire to achieve, and the
confidence in actually achieving it.

When people say that they want "the most" or "the best"
they usually mean the most or best of something of inherent
and substantive value. "I want the best education for my kids so
they can..." "I want to earn as much money as possible so my
family can..." On page 68 we defined substantive optimization
as a procedure that uses assumptions, knowledge, data, rules
of reasoning, and so on, to select a solution or action whose
predicted outcome is best according to a specified criterion of
what is good or valuable. Substantive optimization is a method to
achieve the best possible outcome such as education, happiness,
health, and so on.

A major challenge in substantive optimization is that the
knowledge used to identify the decisions that lead to the optimal
outcome is often wrong in important but unknown ways. This
led us to define, on page 76, a procedural optimum: the unique
solution that optimizes the robustness against surprise while
achieving goals that are substantively adequate. The procedural
optimizer maximizes the robustness against uncertainty and
satisfices the quality of the outcome.[22]

In both cases—outcome optimizing and robust satisficing—
you have to decide what are "goods" (like happiness, that an

[22] To "satisfice" means to "decide on and pursue a course of action that
will satisfy the minimum requirements necessary to achieve a particular
goal." (*Oxford English Dictionary*). We will discuss satisficing extensively in the
next chapter.

outcome optimizer tries to maximize) or "bads" (like crime, that an outcome optimizer tries to minimize). Deciding what is good (What makes you happy?) or what is bad (Is it wrong to smoke marijuana?) is not always easy. Both the outcome optimizer and the robust satisficer must decide what are goods and bads. But the robust satisficer must also decide how much more of a good thing is really necessary to make it all worth while or, equivalently, how much less is the least-acceptable level?

That decision—how much more of a good thing is really needed—is supported in part by recognizing an irrevocable trade off between confidence and quality.

The central idea here is sometimes affectionately called the pessimist's theorem. In its most colloquial form the theorem hides behind the saying "You think things are bad now? Well, they can always get worse." In a more staid and formal setting we begin by recalling Knightian uncertainty and Shackle–Popper indeterminism, discussed in Chapter 3. Uncertainty—the potential for innovation, discovery, and surprise—is unbounded. The realm of the unknown stretches out to a distant horizon that, like the global horizon, recedes before us as we march towards it. As we lift our gaze toward the horizon we capture an ever wider realm of possibilities. And as the horizon of uncertainty increases, the worst that can happen gets worse and worse because more adverse contingencies become possible.[23]

Inverting this we can say that, as our requirements or aspirations get more demanding, we can tolerate less and less uncertainty. If our requirements are very modest, then they will be satisfied anywhere up to some fairly far horizon of uncertainty, regardless of what actually happens. Very modest aspirations

[23] There is of course an analogous optimist's theorem based on the potential for favorable surprises: "You think things are okay? Well, they could be fantastic." This idea is developed in info-gap theory with the opportuneness function and will be discussed later, p. 105.

would be satisfied even if things turn out in any of a multitude of quite different ways from what we expected. But, as our requirements become more demanding, the range of contingencies that do not jeopardize our aspirations begins to contract and we become more vulnerable to adverse contingencies. In other words, as the quality of the required outcome increases, the range of tolerable surprise decreases. As the required outcome becomes more demanding, the vulnerability to uncertainty increases, the robustness against surprise falls off, and our confidence in achieving the required outcome diminishes.

The pessimist's theorem thus implies that the outcome optimizer is actually choosing the decision that is most vulnerable to uncertainty, in the mistaken belief that predicted outcomes are dependable. In contrast, the robust satisficer recognizes the trade-off between how demanding we are in our outcome requirements and how much confidence we can have in achieving those required outcomes. Appreciation of this trade-off assists in deciding how much more of a good thing is really needed: by aspiring to more, we inevitably become less confident in achieving our goal.

This of course does not mean that one cannot or should not adopt high aspirations. Predicted outcomes (whose robustness to uncertainty vanishes) can motivate our thoughts. The pessimist's theorem does, however, imply a need for realism in balancing confidence versus quality.

Let's consider an example.[24] Suppose we hear about an investment opportunity that seems to offer very high return, though it is recognized as being rather risky. Large returns would be great, but we would be satisfied with more modest returns. We have a fixed budget with which to construct an investment portfolio,

[24] Yakov Ben-Haim, 2012, "Doing our best: Optimization and the management of risk", *Risk Analysis*, 32(8): 1326–32.

and we can choose from among a large number of alternative investments.

Usually only one portfolio of investments has maximum predicted return. However, many different portfolios will have the same predicted return at any specified level below this maximum. We could forego the attempt to maximize the return (even on average), and instead choose among many alternative portfolios, all of which are adequate in terms of predicted average return. Specifically, we can choose the portfolio whose predicted return is adequate and that is maximally robust against uncertainty about the returns. We would thereby reduce our requirement for future return in exchange for robustness against surprise. The availability of this robust-satisficing trade-off is an inevitable consequence of uncertainty. The trade-off means that the predicted maximal return is the least robust (most vulnerable) to uncertainty, and that the robustness increases as the outcome requirement is reduced.

In the next chapter we will delve further into the management of innovation dilemmas.

5

Managing Innovation Dilemmas

"'The first thing I've got to do,' said Alice to herself, as she wandered about in the wood, 'is to grow to my right size again; and the second thing is to find my way into that lovely garden. I think that will be the best plan.'

"It sounded an excellent plan, no doubt, and very neatly and simply arranged; the only difficulty was, that she had not the smallest idea how to set about it."[1]

Innovation dilemmas pervade many personal and professional decisions. This chapter explores a method for responsibly handling those dilemmas. The approach is based on two fundamental concepts—robustness and satisficing—that are discussed in the first two sections. We then summarize guidelines for managing innovation dilemmas, and consider several examples: bipolar disorder and pregnancy, rural poverty, military tactics of Epaminondas in ancient Greece, and stock market investments.

Robustness and Locke's Wingless Gentleman

Our ancestors have made decisions under uncertainty ever since they had to stand and fight or run away, eat this root or that berry, sleep in this cave or under that bush. Our species is distinguished

---

[1] Lewis Carroll, *Alice's Adventures in Wonderland*, chapter 4.

The Dilemmas of Wonderland: Decisions in the Age of Innovation. Yakov Ben-Haim.
© Yakov Ben-Haim 2018. Published in 2018 by Oxford University Press.
DOI: 10.1093/oso/9780198822233.001.0001

by the extent of deliberate thought preceding decision. The human ability to decide in the face of the unknown was born from primal necessity.

Betting is one of the oldest ways of deciding under uncertainty. But "you bet you" that "bet" is a subtler concept than one might think, and it will lead us to the concept of robustness.

We all know what it means to make a bet, but just to make sure let's quote the *Oxford English Dictionary* (*OED*): "To stake or wager (a sum of money, etc.) in support of an affirmation or on the issue of a forecast." The word has been around for quite a while. Shakespeare used the verb in 1600: "Iohn a Gaunt loued him well, and betted much money on his head."[2] Drayton used the noun in 1627 (and he wasn't the first): "For a long while it was an euen bet ... Whether proud Warwick, or the Queene should win."

An even bet is a fifty–fifty chance, an equal probability of each of two outcomes. But betting is not always a matter of chance. Sometimes the meaning is just the opposite. According to the *OED*, "You bet" or "You bet you" are slang expressions meaning "be assured, certainly". For instance, Sayers wrote: " 'Can you handle this outfit?' 'You bet,' said the scout."[3] Similarly, Mark Twain reported that "Hank Monk [the driver] said ... 'I'll get you there on time'—and you bet you he did, too."[4]

So "bet" is one of those words whose meaning stretches from one idea all the way to its opposite. Drayton's "even bet" between Warwick and the Queen means that he has no idea who will win. In contrast, Twain's "you bet you" is a statement of certainty. In Twain's or Sayers' usage, it's as though uncertainty combines

---

[2] William Shakespeare, *Henry IV, Pt. 2* III. ii. 44.

[3] D. L. Sayers, *Lord Peter Views the Body*, iv. 68.

[4] Mark Twain, *Roughing It*, University of California Press, Berkeley, CA, chapter 20, p. 132.

with moral conviction to produce a definite resolution. This is a dialectic in which doubt and determination form decisiveness.

John Locke may have had something like this in mind when he wrote:

> If we will disbelieve everything, because we cannot certainly know all things; we shall do muchwhat as wisely as he, who would not use his legs, but sit still and perish, because he had no wings to fly.[5]

The absurdity of Locke's wingless gentleman starving in his chair leads us to believe, and to act, despite our doubts. The moral imperative of survival sweeps aside the paralysis of uncertainty. The consequence of unabated doubt—paralysis—induces doubt's opposite: decisiveness.

But rational creatures must have some method for reasoning around their uncertainties. Locke does not intend for us to simply ignore our ignorance. But if we have no way to place bets—if the odds simply are unknown—then what are we to do? We cannot "sit still and perish".

This is where the strategy of robustness comes in. "Robust" means "Strong and hardy; sturdy; healthy". By implication, something that is robust is "not easily damaged or broken, resilient". A statistical test is robust if it yields "approximately correct results despite the falsity of certain of the assumptions underlying it" or despite errors in the data (*OED*).

A decision is robust if its outcome is satisfactory despite error in the information and understanding that justified or motivated the decision. A robust decision is resilient to surprise, immune to ignorance.

It is no coincidence that the colloquial use of the word "bet" includes concepts of both chance and certainty. A good bet can tolerate large deviation from certainty, large error of information.

---

[5] John Locke, 1706, *An Essay Concerning Human Understanding*, I. i. 5.

A good bet is robust to surprise. "You bet you" does not mean that the world is certain. It means that the outcome is certain to be acceptable, regardless of how the world turns out. The scout will handle the outfit even if there is a rogue in the ranks; Twain will get there on time despite snags and surprises. A good bet is robust to the unknown.

(Even) God is a Satisficer

To "satisfice" means to "decide on and pursue a course of action that will satisfy the minimum requirements necessary to achieve a particular goal" (*Oxford English Dictionary*). Herbert Simon (winner of the 1978 Nobel Prize in Economics) was the first to use the term in this technical sense, which is an old alteration of the ordinary English word "satisfy". Simon wrote "Evidently, organisms adapt well enough to 'satisfice'; they do not, in general, 'optimize'."[6] Agents satisfice, according to Simon, due to limitation of their information, understanding, and cognitive or computational ability. These limitations, which Simon called "bounded rationality", force agents to look for solutions that are good enough, though not necessarily optimal. The optimum may exist, but it cannot be known by the decision maker who has limited information or resources.

There is a deep psychological motivation for satisficing, as Barry Schwartz discusses in *Paradox of Choice: Why More Is Less.* "When people have no choice, life is almost unbearable." But as the number and variety of choices grows, the challenge of deciding "no longer liberates, but debilitates. It might even be said to tyrannize." "It is maximizers who suffer most in a culture that provides too

---

[6] Herbert Simon, 1956, "Rational choice and the structure of the environment", *Psychological Review*, 63(2), 129–38.

many choices"[7] because their expectations cannot be met, they regret missed opportunities, worry about social comparison, and so on. Maximizers may acquire or achieve more than satisficers, but satisficers will tend to be happier.

Psychology is not the only realm in which satisficing finds its place. Satisficing—as a decision strategy—has systemic or structural advantages that suggest its prevalence even in situations where the complexity of the human psyche is irrelevant. We will discuss an example from the behavior of animals.

A prevailing notion in behavioral ecology maintains that animals employ foraging strategies that attempt to maximize energy gain. Animals feeding in a given location deplete its food resources, and thus they tend to move from one feeding site (a "patch" in ecologists' jargon) to another. Optimal foraging theory predicts that an animal is expected to leave a patch as soon as its resources fall below the average level of alternative patches, accounting for travel energy cost and perhaps other considerations.

An ecological colleague of mine at the Technion, Prof. Yohay Carmel, once posed the following question: Why do foraging animals move from one patch to another later than would seem to be suggested by strategies aimed at maximizing caloric intake? Of course, animals have many goals in addition to foraging. They must keep warm (or cool), evade predators, rest, reproduce, and so on. Many mathematical models of foraging by animals attempt to predict "patch residence times" (PRTs): how long the animal stays at one feeding patch before moving to the next one. A common conclusion is that PRTs are under-predicted when the model assumes that the animal tries to maximize caloric

[7] Barry Schwartz, 2004, *Paradox of Choice: Why More Is Less*, Harper Perennial, New York, pp. 2, 225.

intake. Models do exist that "patch up" the PRT paradox, but the quandary still exists.

Yohay and I wrote a paper[8] in which we explored a satisficing—rather than maximizing—model for patch residence time. Here's the idea. The animal needs a critical amount of energy to survive until the next foraging session. More food might be nice, but it's not necessary for survival. The animal's foraging strategy must maximize the confidence in achieving the critical caloric intake. So maximization is taking place, but not maximization of the substantive "good" (calories) but rather maximization of the confidence (or reliability, or likelihood, but these are more technical terms) of meeting the survival requirement. In short, the animal must robustly satisfice its caloric requirement for survival.

We developed a very simple foraging model based on info-gap theory. The model predicts that PRTs for a large number of species—including invertebrates, birds, and mammals—will tend to be longer (and thus more realistic) than predicted by most energy-maximizing models.

This conclusion—that the strategy of robust satisficing seems to predict observed foraging times better than outcome maximizing—is tentative and preliminary (like most scientific conclusions). Nonetheless, it seems to hold a grain of truth, and it suggests an interesting idea. Consider the following syllogism.

1. Evolution tends to select those traits that enhance the likelihood of survival.
2. Animals seem to have evolved strategies for foraging that satisfice (rather than maximize) the energy intake.

[8] Yohay Carmel and Yakov Ben-Haim, 2005, "Info-gap robust-satisficing model of foraging behavior: Do foragers optimize or satisfice?" *The American Naturalist*, 166: 633–41.

3. Hence satisficing seems to be competitively advantageous. Satisficing seems to be a better bet for survival than outcome maximizing.

Unlike the psychologist Barry Schwartz, we are not talking about happiness or emotional satisfaction. We're talking about survival of dung flies or blue jays. It seems that aiming to do good enough, but not necessarily the best possible, is a prevalent aspect of the biological world.

And this brings me to the suggestion that (even) God is a satisficer. The word "good" appears quite early in the Bible: in the fourth verse of the first chapter of Genesis. "And God saw the light [that had just been created], that it was good...". At this point, when the world is just emerging out of *tohu v'vohu* (chaos), we should probably understand the word "good" as a binary category, as distinct from "bad" or "chaotic". The meaning of "good" is subsequently refined through examples in the coming verses. God creates dry land and oceans and sees that they are good (1: 10). Grass and fruit trees are seen to be good (1: 12). The sun and moon are good (1: 16–18). Swarming sea creatures, birds, and beasts are good (1: 20–1, 25).

And now comes a real innovation. God reviews the entire creation and sees that it is *very* good (1: 31). It turns out that goodness comes in degrees; it's not simply binary: good or bad. "Good" requires judgment; ethics is born. But what is particularly significant here is that God's handiwork isn't excellent or optimal. Shouldn't we expect the very best? I'll leave this question to the theologians, but it seems that God is a satisficer.

Some Guidelines

Uncertainty is prevalent in our innovative age, so it makes sense to satisfice our goals while striving for robustness against uncertainty. What does this entail, and how is it done? Like Alice, we know what we want, and now we need to know "how to set about it."

Awareness of the prevalence and profundity of uncertainty, ignorance, and surprise is the first step in managing an innovation dilemma. Beyond that, we can identify processes for thinking through an innovation dilemma and making a choice, based on the ideas of robustness and satisficing. In rough outline:

1. *Set your goals.* What must be achieved in order for the outcome to be acceptable? How good an outcome do you really need? Stated pessimistically, what is the worst outcome that you are willing to accept?

2. *Identify your options and your means.* What alternative lines of action are available? What choices or decisions can be made? What resources are you able, or willing, to devote to achieving your goals?

3. *Identify vulnerabilities to uncertainty.* We have knowledge and understanding that is relevant to the decision at hand, and topical experts can tell us what they know. We cannot know what will be discovered or invented in the future, as discussed in Chapter 3. We can, however, explore the implications of error—especially dramatic and surprising error—in what we currently believe to be true. We can identify vulnerabilities to uncertainty by exploring the implications of error in specific aspects of our current understanding.

4. *Robustify against surprise.* Prioritize the options according to their immunity against uncertainty: an option that achieves your goals over a wider range of uncertainty is preferred over an option that is more vulnerable to surprise. In short: satisfice your goals and optimize your robustness.

5. *Be wary of outcome optimization.* The best outcome is better, in substantive terms, than all others. The problem is that

limited knowledge and unbounded potential for surprise preclude the identification of the option with the best outcome. When facing deep uncertainty, it is better to optimize the robustness against uncertainty for achieving critical goals. This is different from seeking the substantively best outcome. Robust satisficing will usually be preferred over outcome optimizing.

6. *Proxies for robustness.* In some situations we can evaluate the robustness quantitatively: using numerical data and mathematical models.[9] In other situations we must rely on qualitative analysis that is often supported by analyzing conceptual proxies for robustness. We discuss six concepts that overlap significantly with the concept of robustness against uncertainty, and that are useful in the qualitative assessment of decisions under uncertainty. Each of these six concepts emphasizes a different aspect of the overall problem, though they also overlap one another. A decision, policy, action, or system is highly robust against uncertainty if it is strong in most or all of these attributes; it has low robustness if it is weak in all of them. In choosing between two options, the robust preference (if there is one) would be for the option that is stronger in more attributes. The six proxies for robustness are defined as follows.

(i) *Resilience*: rapid recovery of critical functions. Failures and shortfalls are likely, so one should build recovery capability to make one's solutions robust against adverse surprise.

[9] Many examples of quantitative analysis of robustness, based on info-gap theory, have been published. See info-gap.com. See also Yakov Ben-Haim, 2006, *Info-Gap Decision Theory: Decisions Under Severe Uncertainty*, 2nd edn., Academic Press, London. Yakov Ben-Haim, 2010, *Info-Gap Economics: An Operational Introduction*, Palgrave Macmillan, London.

(ii) *Redundancy*: multiple alternative solutions. "Two is better than one,... and the threefold strand will not quickly break" said Solomon.[10] Robustness to surprise can be achieved by having alternative solutions at hand.

(iii) *Flexibility* (sometimes called agility): rapid modification of tools and methods. Agility, as opposed to stodginess, is often useful in recovering from surprise. A physical or organizational system, or a policy, or a decision procedure, is robust to surprise if can be modified in real time, on the fly.

(iv) *Adaptiveness*: adjust goals and methods in the mid to long term. Be willing to adjust as your knowledge changes. The thought process for managing Knightian uncertainty or an innovation dilemma is rarely a once-through procedure. We often have to re-evaluate and revise assessments and decisions. The emphasis is on the longer time range, as distinct from on-the-spot flexibility.

(v) *Margin of safety* (sometimes called preponderance): excess of the benefits (and deficiency of the drawbacks) beyond what is actually required. A margin of safety is not a maximum (or minimum); it is a buffer between adequacy and failure, a predominance in number, quality, or importance of relevant attributes.

(vi) *Comprehensiveness*: interdisciplinary system-wide coherence. The outcome of one's decisions can be impacted by technology, organizational structure and capabilities, cultural attitudes and beliefs, historical context, economic constraints and opportunities, and other factors. Robust decisions will address the multifaceted nature of the problem at hand.

[10] Ecclesiastes, 4: 9–12.

7. *Exploit opportunities.* One may wish to seek out and exploit opportunities, perhaps aiming at better-than-anticipated outcomes. This is different from robust satisficing, as we illustrate on p. 105.

We now consider several examples, illustrating different uses of these conceptual guidelines.

Bipolar Disorder and Pregnancy: Revisited

In Chapter 2 we discussed the innovation dilemma facing the psychiatrically bipolar woman who wants to become pregnant and give birth to a child. Modern medications raise the possibility of achieving the delights of parenthood while also managing the challenges of a severe psychiatric condition. The dilemma is between refraining from pregnancy and foregoing the joys of parenthood, or opting for the promise of new and emerging medications while risking the adverse effects of unforeseen impacts on both mother and child. In Chapter 4 we observed the optimistic idealism expressed in the language of maximizing the benefit and minimizing the risk, while noting the extensive uncertainties that accompanied the decision to be made.

These aspirations for optimal outcomes reflect the patient's sincere desires and the physicians' sense of moral obligation to do their best. However, when dealing with deep uncertainty, we must establish aspirations in light of the irrevocable trade-off between the aspired-to quality of the outcome and the robustness against our ignorance: higher aspirations are more vulnerable to surprise, as discussed on p. 76. To choose an action because our (highly uncertain) understanding predicts that its outcome is best, should be done (if at all) only after acknowledging that this prediction rests on substantial error.

One should of course do one's best. The question is, however, what should one try to optimize: the substantive outcome or the procedural robustness? The robust satisficing approach advocates maximizing immunity against the rampant potential for adverse surprise, while aiming to achieve critical or essential goals. These critical goals may be less than our understanding suggests could be achieved, but they will be acceptable and would constitute success of the decision. Robust satisficing attempts to maximize the confidence in achieving an acceptable outcome, which is different from outcome optimization which attempts to achieve the best predicted result.

We will now explore the robust satisficing approach to thinking through the challenge facing the bipolar woman and her medical team.

The woman, perhaps together with her family, may have any of several **goals**.

One set of goals regards family aspirations. She may wish to give birth to a child. She may wish to give birth to a *healthy* child. She may wish to give birth to more than one child. She may wish to raise children. These goals are diverse and not necessarily mutually contradictory.

Another set of goals regards the woman herself. She may wish to survive the pregnancy without psychiatric catastrophe. She may wish to make it through the pregnancy with no worse than moderate psychiatric destabilization. She may wish to maintain psychiatric stability and comfort throughout the pregnancy.

Whatever the goals, it is important to identify them as clearly as possible. However, even the goals may be uncertain. For instance, the healthy-child goal is uncertain in part because its meaning is unclear. The woman may not appreciate or understand the teratogenic implications, for the child, of the psychiatric medications that will be needed. She may not appreciate the genetic implications, for the child, of her psychiatric condition. Indeed, these implications may be imperfectly known even to medical professionals.

A range of **options** are available for achieving goals from among those we have identified.

Regarding options for the family-oriented goals, one option is pregnancy and birth by the bipolar woman herself (here the goal is itself the only option). Another option is pregnancy of a surrogate mother either from her own egg fertilized by the bipolar woman's partner, or from the implantation in the surrogate mother's uterus of a fertilized egg from the bipolar woman. A third option, addressing the goal of raising children, is adoption.

Regarding psychiatric options, many medications are available and more will likely become available during the pregnancy. Detailed medical advice is needed, throughout the pregnancy, to assess the implications of the options for mother and child.

Whether or not substantial financial means are available is relevant both regarding continual medical and psychiatric assistance and regarding surrogacy.

The **uncertainties** and vulnerabilities to erroneous understanding are substantial, as discussed in Chapter 2. Briefly, the pharmacological management of bipolar disorder is complex, case specific, slow to stabilize, and exceedingly difficult to predict during pregnancy. New medications that become available during the pregnancy are imperfectly understood. The vulnerabilities are lower or higher depending on which specific goals the woman chooses. The goal of "pregnancy and birth without psychiatric catastrophe" entails different uncertainties from "psychiatric stability and comfort with birth of a healthy child". The goal of "raising children" opens the options of surrogacy or adoption that alter the risk profile entirely.

How to achieve **robustness** against uncertainty depends on the goals chosen by the decision maker—Ms. M in the case reported by Dr Burt and her colleagues—and by the available options. Ms. M aspired to pregnancy to full term and birth of a healthy child, together with psychiatric and physiological stability. She also had the financial resources for extensive medical supervision by a team of specialists. Robustness to uncertainty was achieved by the medical team's flexibility in revising the dosage and composition of multiple medications during and after the pregnancy, in response to continuous medical monitoring of mother and fetus.

A different woman (with bipolar disorder) may choose different goals, and thus robustify differently. Imagine an hypothetical woman—call her Ms. H—who aspires first and foremost to raise a gaggle of healthy children. She and her husband would prefer that she bear these children herself, though this is not her primary desire. The uncertainties that she faces arise primarily from side effects of medications during repeated pregnancy. For Ms. H, robustness is maximized by avoiding pregnancy and opting either for surrogacy or adoption. This choice is clear even if Ms. H can afford the pharmacological innovations enabled by continuous surveillance and intervention by a team of medical specialists. Ms. H's goal is most reliably achieved by avoiding the uncertainties of the putative best choice (her own repeated pregnancies).

The comparison of Ms. M with Ms. H shows that the robust optimum depends on the goals, unlike the prediction-based outcome optimum, which is the same for both women. The women may have consulted the same psychiatrist and been presented with the same assessments of risks and opportunities. For both women, the best available evidence (which is evolving and uncertain) indicates that bearing their own children can be safely achieved by continuous medical supervision with a flexible pharmacological regimen. The prediction-based outcome optimum is the same for both women, but the robust optimum is different. Ms. M adopts the innovative pharmacological solution because it enables her primary goal with acceptable risk; Ms. H adopts an alternative option because it most robustly achieves her aspiration.

The robust satisficing decision maker has to make a difficult judgment: What is the goal? What is a good enough or necessary outcome? The prediction-based outcome optimizer is spared this judgment; the unique predicted optimum is specified by the initial conditions and constraints of the problem.[11] However, when facing severe uncertainty, the responsible approach is to aim at essential goals while maximizing one's robustness against the unknowns. This is what both Ms. M and Ms. H have done, each with respect to her own goals. Under deep uncertainty, outcome optimizing is a form of wishful thinking: hoping that our current best understanding is not too far off the mark.

Rural Poverty

Industrial tools for agricultural production, linked with regional or global trade, compatible social attitudes, good goverance, and other factors, can enable high standards of living in rural agricultural populations. Nonetheless, rural poverty persists in many parts of the world. Many diverse factors characterize rural poverty in addition to low agricultural productivity. High mortality or morbidity may be widespread due to lack of medical care. The local population may be suspicious or resentful of the central government or of international non-governmental organizations. Local goverance may differ substantially from the wider norms of the society, and include de facto rule by dominant families, powerful landowners, or warlords.

Technology alone cannot eliminate rural poverty in most cases, though hi-tech innovations can contribute to a solution. Technological innovations can increase agricultural productivity,

[11] In this case the predicted optimum is unique, though sometimes there are multiple optima, all equally good. In any case, what is always unique is the set of equally good options. This set may be empty (no optima), contain a single optimum (the present case), or have multiple equivalent optima.

leading to higher standards of living and less arduous work. New strains of plants, better irrigation methods and infrastructure, better fertilizers and pesticides, mechanization of work, and innovations in many other areas can contribute to reducing and even eliminating poverty.

However, innovative crops may fail when used in environments that differ from where they were developed, causing starvation. Technological innovation can lead to sudden social reorganization and upheaval, thus precluding the purported benefits of the innovation. Increased productivity can lead to population growth, canceling gains in standard of living as more and more mouths must be fed.

In short, technological innovations may be much better than traditional methods of agricultural production, but they may have unforeseen consequences that could make them much worse. How should the well-meaning planner approach this innovation dilemma?

In the year 2000, Prof. Bernard Amadei of the University of Colorado initiated a project in San Pablo, a rural village in the Central American country of Belize. Traditionally, young girls carried water every day from a local river to the village, which prevented these girls from attending school. Amadei recognized that elementary pump technology could provide water while also allowing the girls to go to school with the resulting personal and social benefits of enhanced education. The success of this project, and the lessons learned, led Amadei to found Engineers Without Borders, a voluntary worldwide organization that is devoted to "ethical, environmentally sound and culturally sensitive engineering … to address global challenges".[12]

The water-pump project in San Pablo entails a challenging innovation dilemma. On one hand, machine-powered provision

[12] See http://www.ewb-uk.org, accessed June 26, 2016.

of fresh water and education of children are both clearly desirable. On the other hand, the long-term social and technological
implications are unclear. For example, as water becomes cheaper
and more plentiful, will related infrastructure absorb the new
loads, or could unprocessed sewage become a new health hazard?
How will the education of young girls impact the social status of
women in general, and how will the traditional society respond?
Other challenges can be imagined, and other unanticipated challenges may turn up. The design and implementation of this
promising project needs to address the potential for surprisingly
bad outcomes. We will use a simplified hypothetical reconstruction of the planning of the water-pump project to illustrate the
robustness analysis of this innovation dilemma.

Many engineering projects such as this one are amenable, at
least in part, to quantitative analysis based on numerical data and
mathematical models. However, not all aspects of the problem
can be captured with mathematics. We will focus our example on
the qualitative non-mathematical dimensions of the problem.

We begin by asking three basic questions: What are our goals?
What is our knowledge? What are the uncertainties?

In our hypothetical reconstruction we consider two **goals**:
potable water supplied mechanically, and sustained local support
for the project and its implications. Our **knowledge** includes pro-
ficiency in pump technology and some familiarity with local culture. **Uncertainties** arise in several domains. It is not clear how
long-term technical maintenance of the pump will be achieved,
though it is clear that if the pump system fails after dismantling
the previous water-supply system (water carried by girls) then
catastrophe could result. It is not clear that a stable supply of
fuel for the pump will be available. Finally, our familiarity with
the local culture and tradition is limited, so there are unclear
implications of changing the social status of young girls by removing them from an essential social function (water transport) and

endowing them with longer-term but untested potentials (via education).

Because we cannot quantitatively evaluate the robustness to these uncertainties, we focus on the six qualitative proxies for robustness discussed earlier in this chapter. **Resilience** is the ability for rapid recovery of critical functions after major disruption. **Redundancy** entails multiple alternative solutions. **Flexibility** is the ability for rapid modification of tools and methods. **Adaptiveness** is the ability to adjust goals and methods in the mid to long term, in light of a broad understanding of the problem, as distinct from the rapid response implied by flexibility. **Margin of safety** is a buffer between adequacy and failure. **Comprehensiveness** entails interdisciplinary system-wide coherence.

The engineers will propose various alternative solutions, based on their knowledge, for achieving the stated goals. These alternatives might be based on wood-burning steam-driven pumps, or solar-powered pumps, or perhaps animal-powered rotary systems drawing water through a system of conveyor belts and buckets. Engineers are an inventive lot. Each technical solution can be evaluated for its robustness against uncertainty while achieving the specified goals.

First of all it is critical to **satisfice** the goals. Don't try to maximize or optimize the outcome. The optimizer's fallacy is to exploit all of the available knowledge to select the design that is predicted to have the best substantive outcome (some combination of maximum water and maximum local support). The problem is that part of our knowledge is, in fact, wrong, and by attempting to optimize the outcome we inadvertently exploit our errors along with our correct understanding. Seek an adequate but probably suboptimal solution. Suboptimal solutions are more numerous than solutions that purport to optimize the outcome. By considering all acceptable solutions we have more opportunity to consider designs that augment the robustness against uncertainty while satisfying the design goals.

The engineers are the technical experts, but the challenge is not only technical, as Engineers Without Borders emphasizes. Major uncertainty can arise from incomplete understanding (by the engineers) of local customs and values. The strategy of "co-design" involves local people and their expertise in all stages of the planning and implementation.[13] Co-design by local representatives together with technical experts contributes to **comprehensive** system-wide thinking and enhances local buy-in and support for the project.

The technical solution should have a **margin of safety**: providing more water-carrying capacity than is initially viewed as essential. This will ameliorate local concerns about deficiency of supply, and will enable response to enhanced use that may result from easy accessibility.

It is important to train local people in maintenance and repair of the pump and associated pipes. This provides **resilient** response to the inevitable failures that will arise through ordinary (and some extra-ordinary) use. It also provides some **flexibility** in using the new system by enabling local people to modify and improve the system as new difficulties or opportunities arise.

It may be desirable to consider a transition period of dual supply: pumps providing most of the water, young girls carrying the remainder. This provides **redundancy** both in the water supply and in the role of the girls until technical glitches are worked out and the school system is expanded to accommodate the girls.

[13] The concept of co-design has found wide application in developing successful security strategies initiated by Western powers in non-Western or developing societies. See David Kilcullen, 2013, *Out of the Mountains: The Coming Age of the Urban Guerrilla*, Hurst & Company, London.

Mechanisms for long-term contact with outside expertise provide advice and support and enhance **adaptiveness** by enabling adjustments as context and requirements change over time.

The planner's task is to prioritize the design proposals put forward by the engineers. A proposal that is more robust to uncertainty, while achieving the specified goals, is preferred over a proposal that is less robust and more vulnerable to failure due to unanticipated contingencies. The six proxies for robustness that we have considered—resilience, redundancy, flexibility, adaptiveness, margin of safety, and comprehensiveness—provide qualitative concepts for evaluating and ranking design proposals. The next two sections explore further examples.

Epaminondas's Feint

We will use the Theban–Spartan Wars of the fourth century BCE to illustrate the method of robustly satisfying one's outcome goals.[14] Keegan describes the situation as follows.[15]

> Thebes won two remarkable victories, at Leuctra in 371 and Mantinea in 362, where its outstanding general, Epaminondas, demonstrated that the phalanx system could be adapted to achieve decisive tactical manoeuvre in the face of the enemy. At Leuctra, outnumbered 11,000 to 6,000, he quadrupled the strength of his left wing and, masking his weakness on the right, led his massed column in a charge. Expecting the battle to develop in normal phalanx style, when both sides met in equal strength along the whole front of engagement, the Spartans failed to reinforce the threatened section in time and were broken, for considerable loss to themselves and almost none to the Thebans. Despite this

[14] Yakov Ben-Haim, 2015, "Dealing with uncertainty in strategic decision-making", *Parameters*, US Army War College Quarterly, 45(3) Autumn 2015.

[15] John Keegan, 1994, *A History of Warfare*, Pimlico, London, p. 258. See also John David Lewis, 2010, *Nothing Less than Victory: Decisive Wars and the Lessons of History*, Princeton University Press, Princeton, NJ, p. 52.

warning, they allowed themselves to be surprised in exactly the same fashion at Mantinea nine years later and were again defeated.

A Spartan robust satisficing analysis would begin by identifying the Spartan goal. Given the balance of force favoring the Spartans nearly 2 to 1, the goal could reasonably have been routing the Thebans.

One then outlines the relevant knowledge. This would include intelligence about enemy strength, plans of battle, weapons and tactics, weather, terrain, and so on.

One then identifies the domains of uncertainty, which can be numerous. How confident are we in the intelligence about enemy strength? Might enemy allies be lurking in the region? Is the intended field of battle truly flat and unimpeded? And more.

These three components—the goal, the knowledge, and the uncertainties—are then combined in assessing the robustness to error or surprise of any proposed Spartan plan of battle. This is not a simple task (hindsight is a tremendous aid). The analysis of a proposed decision centers on the "robustness question" which is: how large an error or surprise can the proposed plan tolerate without falling short of the goal? The question being asked is not "How wrong are we?", but rather "How large an error can we tolerate?" These are very different questions, and only the second question is answerable with our current knowledge. Further-more, the question is not "What is the best possible outcome?", but rather "What is the most robust plan for achieving our goals?" These questions also differ fundamentally, and the latter is far more relevant when facing strategic uncertainty.

We won't perform the robustness analysis on all the dimensions of uncertainty. We will focus on the Spartan uncertainty about Theban tactics. The standard tactical model, as Keegan explained, was uniform frontal assault of phalanxes leading to close fighting with swords and spears. The robustness question for the Spartans is: how large a Theban deviation from this combat model would deny Spartan victory? If the Spartans were confident that a two-to-one force ratio was sufficient for victory, then a local two-to-one Theban force concentration entails significant Spartan vulnerability. Given the overall Spartan force advantage, a robust tactic for the Spartans would be to hold significant reserve to either bolster Spartan forces against Theban concentration or to exploit points of Theban weakness.

Holding forces in reserve was not standard operating procedure in those times, and the Spartan general would have viewed it as a new and innovative battle plan. The Spartans, if they had considered this option, would have faced an innovation dilemma. On one hand, the force-reserve option has the potential for routing the Theban forces and even for inflicting devastating losses. On the other hand, this new tactic entails unknowns such as how the Spartan rank and file would react to such unconventional behavior by the commander. The Spartans chose conventional tactics, and paid the price of Epaminondas's feint.

The point of this example is not that holding forces in reserve is a good tactic. The point is the type of reasoning: identify goals, knowledge, and uncertainties, and then maximize one's robustness against surprise. Don't ask for the best outcome; ask for the best robustness in achieving specified outcomes. One is optimizing something (the robustness) but not what is often the aim of optimization (the substantive outcome).

Strategic uncertainty motivates the robust satisficing methodology: optimize one's immunity against surprise, rather than trying to optimize the quality of the outcome. Routing the Thebans on the day of battle is less than a Spartan general might desire. The latter might entail totally destroying their force, their will to fight, their allies' support, the economic base of their future resistance, etc. Routing the Thebans, we suppose in this example, would constitute "success" or "victory" or at least be "good enough", and the aim of the robust satisficing analysis is to achieve this outcome as reliably as possible. What one optimizes is the reliability of a good enough outcome (which can be chosen as ambitiously or as modestly as one wants).

The analysis would continue by examining the vulnerability to additional uncertainties and the robustness obtained from alternative plans of battle. The analysis is neither simple, nor fast, nor free of the need for deliberation and judgment in managing innovation dilemmas. However, the process identifies a plan that will achieve the specified goals over the widest range of surprise by the adversary and error in our knowledge.

Stock Markets

Stock markets provide a venue for investors, of all sorts, to buy and sell securities and to make (and lose) money over long and short durations. In the New York Stock Exchange on a typical August day in 2016, 820 million securities were sold in 2.8 million transactions for a total monetary exchange of $29 billion.[16]

Many of the securities that are traded are fairly reliable investments in stable commercial ventures, and have modest returns. For instance, S&P (Standard & Poor's) 500 stocks have a long-term average annual return of 7%. Other securities are less characteristic of the name "securities" and offer annual (or shorter-term) returns that range from large loss to large gain. For instance, the largest losses or gains on a *single day* can equal or exceed the S&P 500 average *yearly* gain.[17]

Consider a cautious and patient investor whose goal is to achieve reliable long-term growth of wealth for retirement. The

[16] See http://www.nyxdata.com/Data-Products/NYSE-Volume-Summary# summaries, accessed August 15, 2016.

[17] See https://en.wikipedia.org/wiki/List_of_largest_daily_changes_in_the_ Dow_Jones_Industrial_Average, accessed August 15, 2016.

investor feels that having, say, a million dollars would provide a good quality of life in retirement thirty years hence. Having specified the goal, at least approximately, the investor then must decide how much to invest each year and must choose a portfolio of stocks that will achieve the goal despite unknown future developments in the investor's needs, the stock market, the wider economy, and beyond. This is a standard robust satisficing approach, though it's not always recognized as such (brokers or investment firms may present their offerings with different rhetoric, as discussed on p. 67).

Our cautious investor sees a wide range of stock offerings, where stocks with higher returns are typically more uncertain. Venture capital investments, for instance, may yield huge returns, or may fall flat, while S&P 500 or similar portfolios are more modest but quite reliable over the long term. For example, investing $15,000 annually for thirty years in S&P 500 will satisfy the million-dollar goal even if the long-term annual return falls by two percentage points (which is quite unlikely). Having said that, the same annual investment in a speculative venture capital portfolio with projected annual returns of 20% will reach a million dollars after only fifteen years and will achieve much more after thirty years.[18] However, the speculative option is highly uncertain and total loss is a definite possibility. The cautious investor faces an innovation dilemma: what looks like a very promising investment may turn out much worse than the solid familiar option because little is knowable about the promising future trajectory. Conceptual as well as quantitative tools are available for analyzing the options and choosing a portfolio that robustly satisfies the cautious investor's needs and inclinations.[19]

---

[18] These calculations are in "real" dollars, that is, correcting for inflation.

[19] Yakov Ben-Haim, 2010, *Info-Gap Economics: An Operational Introduction*, Palgrave Macmillan, New York.

Now consider a different investor, a speculator whose aspiration is to exploit those wonderful opportunities we're always hearing about. Uncertainty, for this investor, is the opportunity for a windfall, an outcome far better than the norm. This investor is not interested in robustly protecting against the vicissitudes of market turbulence. This investor wants to exploit that turbulence; the unknown future is an opportunity, not a threat; uncertainty is propitious, not pernicious.

This second investor is not incautious or impetuous, but his goals are different from those of the first investor. The speculator thrives on uncertainty but also needs a method for prioritizing the many enticing options. One approach is to ask the "opportuneness question", which is the analog of the robustness question discussed on pp. 75 and 101. Namely, for a given portfolio of stocks, what is the lowest horizon of uncertainty at which this portfolio could yield a rate of return far beyond the expectation? The portfolio is highly opportune if small surprises could result in large benefits. A windfall is an unexpected reward, and the portfolio is highly opportune if windfalls could occur under nearly normal conditions. The opportuneness of the portfolio is a measure of its propensity for exploiting favorable surprise. This is the inverse of the robustness of the portfolio, which assesses the immunity of the portfolio against adverse surprise. The opportune windfaller employs a strategy that is the inverse of the robust satisficer's strategy.

The opportuneness of the portfolio displays a trade-off that is analogous to the trade-off of robustness against required outcome (see p. 76). The opportune windfaller aspires to wonderful outcomes beyond the norm. Greater aspirations (e.g. for greater profit) are possible only at greater uncertainty for which even more favorable surprises are possible. As the opportune windfaller's aspiration increases, the horizon of uncertainty that must be countenanced increases as well. In contrast, as the robust satisficer's critical requirement increases, the horizon of uncertainty that can be tolerated decreases, meaning that the robustness against uncertainty decreases. These trade-offs are in opposite directions, reflecting the complementarity between opportuneness and robustness.

The opportune windfaller, for instance the stock speculator, may also face an analog of the robust satisficer's innovation dilemma. Consider two portfolios of stocks. At low uncertainty, the first portfolio could possibly yield higher-than-expected returns while the second portfolio could perhaps yield far greater returns. In other words, the second portfolio seems more opportune because, at low uncertainty, it has greater potential for higher-than-expected returns than the first portfolio. However, this second portfolio, while quite uncertain (which is good) is nonetheless more familiar and less uncertain, and perhaps less audacious and innovative, than the first portfolio. This means that, as the horizon of uncertainty expands, the excess profit potential of the more-familiar second portfolio grows more slowly than the excess profit potential of the first portfolio. Consequently, there is a level of uncertainty at which the potential for wonderful windfall is greater for the first portfolio. If the investor's aspiration (not to say avarice) for greater-than-anticipated profit is very large, then the first portfolio is more opportune. However, more moderate (not to say modest) aspirations for windfall would favor the second portfolio. A reversal of preference between the portfolios may occur, depending on the investor's aspiration for windfall.

The meanings of the trade-offs for opportuneness and robustness are different, but both entail a dilemma, each in its own domain.

The decision methodologies of robust satisficing and opportune windfalling are conceptual opposites. Nonetheless, the robust satisficer and the opportune windfaller can cooperate for mutual benefit. Consider two investors, Roberta who is a robust satisficer, and Owen who is an opportune windfaller. Roberta owns stocks in a firm that, in her eyes, has deteriorated by moving in adventurous and highly uncertain directions. Roberta is happy to sell; Owen is happy to buy. Their different but complementary attitudes towards uncertainty, and their different goals in investing, make them good trading partners.

6

Cultures of Innovation and Progress

"'Well, in *our* country,' said Alice, still panting a little, 'you'd generally get to somewhere else—if you ran very fast for a long time, as we've been doing.'

"'A slow sort of country!' said the Queen. 'Now, *here*, you see, it takes all the running *you* can do, to keep in the same place. If you want to get somewhere else, you must run at least twice as fast as that!'"[1]

Some innovations are revolutionary, startling, or unsettling, while some innovations are so gradual and evolutionary that we detect them only in retrospect after they have altered our lives and ways of thought. The modes of innovation are numerous, but disputation—in the broad sense of intellectual confrontation and verbal conflict between different attitudes, inclinations, or beliefs—is often a central element in the initiation and adoption or rejection of an innovation. Innovations, like disputes, are often the work of individuals, and in Western society a strong traditional emphasis on individualism has played a major role in encouraging innovation. Much effort is devoted to educating creative, innovative, individualistic citizens. Modern education emphasizes

---

[1] Lewis Carroll, *Through the Looking-Glass*, chapter 2.

The Dilemmas of Wonderland: Decisions in the Age of Innovation. Yakov Ben-Haim.
© Yakov Ben-Haim 2018. Published in 2018 by Oxford University Press.
DOI: 10.1093/oso/9780198822233.001.0001

mastering and retaining extant knowledge, but also focuses on developing the skills and propensities for discovering new truths and identifying old falsehoods. Education for independence of thought is a hallmark of innovative societies. However, seeking universal truths may entail a paradox that is removed only by limiting the universalism itself. The search for universal truths entails innovation dilemmas that encumber all efforts at progress and introduce unavoidable fragilities to innovative societies. Two examples illustrate this: the UN Universal Declaration of Human Rights, and military geopolitical strategy.

Ridiculous, Immoral, Impossible

The mathematician G. H. Hardy is reported to have said: "It is never worth a first class man's time to express a majority opinion. By definition, there are plenty of others to do that."[2] Innovative ideas never start as majority opinions and they are sometimes viewed as absurd, immoral, or impossible, but they can have revolutionary impact, eventually becoming commonplace.

An example of a global change resulting from a specific discovery by a small number of people is provided by the spice trade. From ancient to modern times innumerable ships brought spices from the Far East across the Indian Ocean and up the Red Sea, where countless camels trudged across north-eastern Africa to the Mediterranean Sea where ships carried the spices to Europe. The route was controlled, and taxed, in various times and regions by Indians, Ethiopians, Arabs, and Turks. This was a direct and logical east–west route, and it worked. However, men in the Age of Discovery thirsted for something new and were not inhibited by conventional geographical wisdom. In the last decades of the

---

[2] Attributed to G. H. Hardy in C. P. Snow, 1996, *Variety of Men*, Charles Scribner's Sons, New York, p. 52.

fifteenth century, Portuguese explorer Vasco da Gama, and others, pioneered a sea route around the southern tip of Africa, up the Indian Ocean and on to the spice markets of India and China. This route went south and east before turning north. World trade and the global economy were transformed. A few men, driven by curiosity, ambition, and greed, changed everybody's mental map of the world.

The famous fifth postulate of Euclid's geometry provides another example of an impossibility becoming a truism. The postulate states that, given a line and a point not on that line, there exists one and only one straight line passing through the point and never crossing the line.[3] Generations of mathematicians, from Euclid's day until the nineteenth century, attempted to derive the fifth postulate from the previous four.[4] In the early decades of the nineteenth century János Bolyai and Nikolai Lobachevsky independently showed that this cannot be done.[5] They demonstrated "non-Euclidean" geometries that are just as self-consistent as Euclid's geometry, while violating Euclid's famous fifth. Non-Euclidean geometries developed into a major area of mathematical research, with applications in spherical geometry and navigation, relativistic cosmology, and beyond. Kant—who was unaware of the work of Lobachevsky and Bolyai—mistakenly thought that Euclidean geometry is a priori synthetic: not derived from our senses, nor deduced logically, but still true of the real world.[6] Kant's conviction implied that any assertion contrary to Euclid's postulate would be logically inconsistent and physically impossible. The development of

---

[3] Euclid's original statement is different but logically equivalent. See Thomas Heath, ed., 1925, *The Thirteen Books of Euclid's Elements*, 2nd edn., Cambridge University Press, Dover Publications reprint, New York, 1956, vol. 1, p. 202.

[4] Heath, *The Thirteen Books of Euclid's Elements*, vol. 1, pp. 204–19.

[5] Morris Kline, 1972, *Mathematical Thought From Ancient to Modern Times*, Oxford University Press, Oxford, vol. 3, chapter 36.

[6] Kline, *Mathematical Thought From Ancient to Modern Times*, vol. 3, p. 862.

non-Euclidean geometries overturned this view and contributed to unseating the very idea of unique and absolute truth.

Kuhn's concept of "paradigm breakage"[7] describes abrupt revolutionary change in scientific understanding, and provides additional examples of innovative ideas starting at the fringe and becoming standard doctrine. For example, Einstein's theory of special relativity transformed the meaning of the fundamental physical notions of speed, time, and simultaneity.[8] Consider any two events. For instance, the first piano key stroke in Elton John's concert before the Queen, and the click of the ticket punch of a tardy fan. The gate keeper may rightly assert that these events are simultaneous, while an airborne police officer may rightly disagree. The timing of events depends on the speed and direction of motion of one's frame of reference. Special relativity overturns the conventional idea of universal and absolute time for events occurring at different locations in space. Ideas of moral, political, and even physical relativity did not originate with Einstein. Nonetheless, his revolutionary scientific innovation greatly strengthened the now common belief that "everything is relative".

Ridiculous, immoral, or impossible ideas are usually just that. However, revolutionary innovations most often start out looking like their despised impossible cousins, only to become widely accepted truths, and sometimes even platitudes, in due course.

Truth and Glaciers …

Innovation and the emergence of truth is, however, sometimes gradual and not revolutionary at all. John Kenneth Galbraith wrote that "The movement of ideas toward truth may be glacial

[7] Thomas S. Kuhn, 1970, *The Structure of Scientific Revolutions*, 2nd edn., University of Chicago Press.

[8] A. Einstein, 1905, "Zur Elektrodynamik bewegter Körper" ("On the Electrodynamics of Moving Bodies"), *Annalen der Physik*, (ser. 4), vol. 17, pp. 891–921.

but, like a glacier, it is hard to stop."[9] Much human progress, and many profound truths, evolve through small steps moving innocuously toward something new and grand. Large but gradual change is hardly noticed at any individual step. No dramatic transitions occur, and no long-term plans are ever made, but the cumulative effect is far greater than anticipated from the individual steps. The innovation, if it's even correct to call it that, appears only in retrospect, like the mountain climber who looks back after hours of toil to see the distant valley far below.

For example, David Landes discusses a "revolution in time" that accompanied the invention of the mechanical clock in twelfth- to fourteenth-century Europe.[10] Large clocks in church towers were originally intended to wake the faithful for prayer. However, their central location made them suitable for coordinating other daily activities. Gradually, as clocks became smaller, cheaper, and more widely distributed and reliable, a new attitude towards the passage of time emerged. People developed a sense of time discipline; a moral obligation not to waste or "kill" time. The concept of efficiency—product per unit of time—depends on time discipline. As opposed to a "mañana" attitude of "easy-going procrastination"[11] because "tomorrow" is infinite, time discipline takes its subjects to task now; tomorrow is there for planning, not for putting off what can be done today. This is a dramatic change in attitude, but it was gradual, evolutionary, and unplanned.

Karl Popper's "piecemeal social engineering" is a process of gradual collective improvement:

[9] John Kenneth Galbraith, 1986, *The New Industrial State*, 4th edn., Mentor Books, Penguin Group, New York, p. 201.

[10] David S. Landes, 1983, *Revolution in Time: Clocks and the Making of the Modern World*, Harvard University Press, Cambridge, MA.

[11] *Oxford English Dictionary*, accessed online September 4, 2013.

The characteristic approach of the piecemeal engineer is this. Even though he may perhaps cherish some ideas which concern society "as a whole"—its general welfare, perhaps—he does not believe in the method of re-designing it as a whole. Whatever his ends, he tries to achieve them by small adjustments and re-adjustments which can be continually improved upon.[12]

An example of piecemeal engineering is Herman Kahn's treatise *On Thermonuclear War* which "concentrate[s] on the problem of avoiding disaster and buying time, without specifying the use of this time." Kahn has no prescription for long-term or global solutions to the challenges of nuclear weapons. He analyzes "some of the practical military alternatives" for the US to pursue in the fifteen years following his study.[13] Numerous scholars, soldiers, politicians, journalists, and many others around the world contributed to the gradual understanding and acceptance of a practical modus vivendi that has accompanied the non-use of nuclear weapons in the decades, and then generations, since their first use in 1945.

Kuhn's concept of "normal science"[14] describes the gradual collective progress of scientific knowledge that fills the long intervals between revolutionary paradigm breaks discussed earlier. For example, the quantum revolution in physics (a Kuhnian paradigm shift of gigantic proportions) occurred in the early decades of the twentieth century. Since then physics has evolved by accretion of new, interesting, and important—but not revolutionary—discoveries. Physicists understand, far better than before, radioactive decay of nuclei, laser production of light, semiconduction in solids used in transistors, and many

[12] Karl R. Popper, 1961, *The Poverty of Historicism*, Harper Torchbooks, New York, p. 66.

[13] Herman Kahn, 1961, *On Thermonuclear War*, 2nd edn., Princeton University Press, Princeton, NJ, p. 7.

[14] Kuhn, *The Structure of Scientific Revolutions*, 2nd edn.

other phenomena. The development of the bubble chamber and of high-energy particle accelerators led to the discovery of a multitude of subatomic particles. Scientific papers have been written and Nobel Prizes have been awarded. New gauge theories have emerged, and a plethora of string theories have evolved and been vigorously debated.[15] But this is all normal science. The next Kuhnian paradigm shift remains nestled in the future. We cannot know which of today's disputes will lead to tomorrow's paradigm shift. Today's dispute is often the kernel of tomorrow's progress.

Cultures of Disputation

Disputation is a verbal process of controversy through which ideas are explored, truths are tested, and progress is made (or lost). The parties to a dispute express views, arguments, emotions, perhaps even threats. A dispute may degenerate into a violent physical conflict, but a dispute, as such, is not violent. The dispute is resolved if some (or all) of the parties change their positions and agree on a common stand. Or a dispute is resolved or abated by compromise without full agreement. Or a dispute may remain unresolved. There are diverse cultures of disputation, some more beneficial than others.

An unresolved dispute may simply be forgotten, perhaps because it becomes irrelevant. Or, a dispute may remain unresolved between the disputants, but other people or the broader society may resolve the dispute either by adopting a compromise or by adopting one side or the other, or perhaps adopting another position altogether. Or, an unresolved dispute may lead to factiousness and fragmentation of society.

Disputation can be dangerous or counterproductive if it is exploited by rhetorical skill to sway or deceive, as Socrates

---

[15] Lee Smolin, 2006, *The Trouble with Physics: The Rise of String Theory, the Fall of a Science and What Comes Next*, Penguin Books, London.

warned Phaedrus. By way of example, Socrates suggested that a soldier should ride into battle on "that tame animal which possesses the largest ears" (a donkey rather than on a horse) with arguments that Phaedrus deemed "utterly ridiculous". Socrates responds that

> when a master of oratory, who is ignorant of good and evil, employs his power of persuasion on a community as ignorant as himself, not by extolling a miserable donkey as being really a horse, but by extolling evil as being really good, and when by studying the beliefs of the masses he persuades them to do evil instead of good, what kind of crop do you think his oratory is likely to reap from the seed thus sown?[16]

The gist of Socrates' (rhetorical) question is that disputation for the sole purpose of persuasion, without an eye to removing an uncertainty or discovering a truth, may be detrimental.

The culture of disputation can beneficially nurture the growth and diversification of groups, organizations, or society. Disputation can encourage the critical examination of ideas and preconceptions, leading to the sprouting and nurturing of new lines of thought. Disputation can support the growth of what Huntington, in discussing civil–military relations, called "strategic pluralism: the multiplication of programs and activities and a tendency to equalize the division of resources among competing military claims."[17]

Huntington explains that disputation is a responsibility of leaders:

> Too much harmony is just as much a symptom of bad organization as too much conflict. On the face of it, something is

[16] Plato, *Phaedrus*, 260b,c, in *The Collected Dialogues of Plato, including the Letters.* Edith Hamilton and Huntington Cairns, editors, Princeton University Press, Princeton, NJ, 1961.

[17] Samuel P. Huntington, 1957, *The Soldier and the State: The Theory and Politics of Civil–Military Relations*, Harvard University Press, Cambridge, MA, p. 421.

wrong with a system in which, during the course of a four-year major war, the political Chief Executive only twice overrules his professional military advisers. This can only mean that one of them was neglecting his proper function and duplicating the work of the other.[18]

The Talmud is prototypical of a fruitful culture of disputation. Hillel and Shamai had many disputes. For example, they did not agree about the meaning of the Biblical command to be fruitful and to multiply.[19] Hillel claimed that the commandment is fulfilled if one has no less than one son and one daughter. Shamai claimed: no less than two sons. Hillel and Shamai resolved few of their numerous disputes. However, perhaps their greatest contribution was to legitimize the culture of disputation that they embodied.

Without disputation, errors will occur that could otherwise have been avoided, and progress will be slow or lacking. Disputation in the search for innovation and discovery will be beneficial, while cultures of contrariness or deceit will suppress the individual's drive to discovery.

Individualism and Progress

Social stagnation can occur for many reasons, just as there are many different formulas for innovation and progress. Progress in Western countries has emerged along various paths, but an element of aggressive (sometimes even cut-throat) individualism is characteristic, along with other mores and institutions.

Consider the case of Dr Thomas Stockmann in Ibsen's play *An Enemy of the People.* Dr Stockmann was the medical officer at the new municipal baths of his town. He discovered a pernicious

[18] Huntington, *The Soldier and the State*, p. 329.
[19] Genesis 1: 28.

contamination in the bath waters and insisted that the baths be closed. The community leaders rejected his advice on economic grounds, eliciting an angry response from Dr Stockmann:

> The majority is never right. Never, I tell you! That's one of these lies in society that no free and intelligent man can help rebelling against. Who are the people that make up the biggest proportion of the population—the intelligent ones or the fools?

Dr Stockmann was right on the medical issue but he may have exaggerated his social claim. Nonetheless, there is some truth in his exclamation. Consider the distribution of any trait in a large population, for example body height. Some people are tall and some short, and most are somewhere in the middle range. Nonetheless, we can always divide a large population into two equal halves: those who are taller than the median height and those who are shorter. Half of any large population is taller than the other half.

This is true also of other more important traits, even when they are difficult to measure. Half of any population has higher IQ than the other half. Half of any population is more courageous (by some criterion) than the other half. And so on.

But just as we can divide a population fifty-fifty on any trait, we can also divide it at different ratios. One percent of any large population has higher IQ, or greater courage, avarice, stubbornness, or curiosity, than the remaining 99 percent.

Identifying the traits that enable major innovation is beyond our scope. But there is no doubt that most innovators display some—usually several—rare attributes. Innovators have star qualities that enable them to see possibilities that others miss, and to bring these visions to fruition. Nurturing individual distinctiveness will surely enhance innovation and progress.

But individualism is not enough to support an innovative society. An individual's pursuit of goals, visions, and interests can lead to innovation and progress, but this is predicated on freedom from other worries. In 1932 Bertrand Russell argued that, as a result of industrial mechanization, a four-hour work day can provide all the requirements for stable and happy life. After four hours of work one has many hours for free pursuit of one's interests.

Russell also argued that, throughout history, a leisure class

cultivated the arts and discovered the sciences; it wrote the books, invented the philosophies, and refined social relations. Even the liberation of the oppressed has usually been inaugurated from above. Without the leisure class, mankind would never have emerged from barbarism.[20]

Industrialization, Russell claimed, frees everyone to enjoy an amount of leisure that historically was limited to small segments of society.

Nonetheless, Russell recognized that leisure will lead to fruitful innovation in only a small fraction of society:

At least one percent will probably devote the time not spent in professional work to pursuits of some public importance, and, since they will not depend upon these pursuits for their livelihood, their originality will be unhampered, and there will be no need to conform to the standards set by elderly pundits.[21]

One percent is not much, but it could be very significant if it is one percent of the entire population. How to encourage and nurture this innovative one percent? "The wise use of leisure", Russell declares, "is a product of civilisation and education."[22] Education would seem to be the only sensible route to instilling

[20] Bertrand Russell, "In Praise of Idleness", 1932, appearing in *In Praise of Idleness and Other Essays*, Unwin Paperbacks, London, 1960, p. 23.

[21] Ibid. p. 25.

[22] Ibid. p. 18.

those values and ideals that can induce innovation and lead us from barbarism. Russell's vision depends on education for all. How can education that is open to everyone and caters to all, from preschool through university, nurture those rare few who tower above the norm? How can massive public education nurture an individualistic innovative elite? Western societies have grappled with this challenge since the early nineteenth century. Education is ultimately an individual process, especially when we aim to cultivate exceptional individuals. Much effort has been devoted to improving educational methods, and modern computer technology opens new possibilities. We now consider one of them.

MOOCs: Massive Open Online Courses

MOOCs—massive open online courses—have fed millions of knowledge-hungry people around the globe. Stanford University's MOOCs program has taught open online courses to tens of thousands of students per course, and has had millions of enrollees from nearly every country in the world.[23] The students hear a lecturer, and also interact with each other in digital social networks that facilitate their mastery of the material and their integration into global communities of the knowledgable. The Internet, and its MOOC realizations, extend the democratization of knowledge to a scale unimagined by early pioneers of workers' study groups or public universities. MOOCs open the market of ideas and knowledge to everyone, offering everything from esoteric spirituality to practical computer languages. It's all there; all you need is a browser and an Internet connection. The questions we ask are: Can MOOCs nurture Russell's innovative one percent? What sort of innovation dilemma do we face here?

[23] Many other universities also offer MOOCs. My own university, for example, has many MOOCs, including the first MOOC offered in Arabic anywhere in the world.

The Internet is a facilitating technology, like the invention of writing or the printing press, and its impacts may be as revolutionary. MOOCs are here to stay, like the sun to govern by day and the moon by night, and we can see that it is good. But MOOCs also have limitations or perhaps even dangers, some of which we can begin to understand, while others are still hidden in the future. This entails an innovation dilemma bearing on the relation between individualism and innovation-based progress.

Education depends on the creation and transfer of knowledge. Insight, invention, and discovery underlie the creation of knowledge, and they must precede the transfer of knowledge. MOOCs enable learners, anywhere and anytime, to sit at the feet of the world's greatest creators of knowledge.

The distinction between creation and transfer of knowledge is necessarily blurred in the process of education itself. Deep and meaningful education is the creation of knowledge in the mind of the learner. Education is not the transfer of digital bits between electronic storage devices. Education is the creation or discovery by the learner of thoughts that previously did not exist in his or her mind. One can transfer facts per se, but if this is done without creative insight by the learner it is no more than Huck Finn's learning "the multiplication table up to six times seven is thirty-five".[24]

Invention, discovery, and creation occur in the realm of the unknown; we cannot know what will be created until it appears. Two central unknowns dominate the process of education, one in the teacher's mind and one in the student's.

The teacher cannot know what questions the student will ask. Past experience is a guide, but the universe of possible questions is unbounded, and the better the student, the more unpredictable

<hr/>

[24] Mark Twain, *The Adventures of Huckleberry Finn*, Harper & Brothers, New York and London, chapter 4.

the questions. The teacher should respond to these questions because they are the fruitful meristem of the student's growing understanding. The student's questions are the teacher's guide into the student's mind. Without them the teacher can only guess how to reach the student. The most effective teacher will personalize his interaction with the student by responding to the student's questions.

The student cannot know the substance of what the teacher will teach; that's precisely why the student has come to the teacher. In extreme cases—of really deep and mind-altering learning—the student will not even understand the teacher's words until they are repeated again and again in new and different ways. The meanings of words come from context. A word means one thing and not another because we use that word in this way and not that. The student gropes to find out how the teacher uses words, concepts, and tools of thought. The most effective learning occurs when the student can connect the new meanings to his own existing and growing mental contexts. The student cannot always know what contexts will be evoked by his learning.

As an interim summary we can say that learning can take place only if there is a gap of knowledge between teacher and student. This knowledge gap induces uncertainties on both sides. Effective teaching and learning occur by personalized interaction to dispel these uncertainties, to fill the gap, and to complete the transfer of knowledge.

We can now appreciate the most serious pedagogic limitation of MOOCs as a tool for education, especially for nurturing exceptional individuals, and perhaps point to a solution.

Mass education is democratic, and MOOCs are far more democratic than any previous mode. However, democracy in education creates a basic tension. The more democratic a mode of communication, the less personalized it is because of its massiveness. The less personalized a communication, the less effective it is pedagogically, especially in nurturing rare individualistic innovativeness. The gap of the unknown that separates teacher and learner is greatest in massively democratic education.

Socrates inveighed against the writing of books.[25] They are too impersonal and immutable. They offer too little room for Socratic midwifery of wisdom, in which knowledge comes from dialog. Socrates wanted to touch his students' souls, and because each soul is unique, no book can bridge the gap. Books can at best jog the memory of learners who have already been enlightened. Socrates would probably not have liked MOOCs either, and for similar reasons.

Nonetheless, Socrates might have preferred MOOCs over books because the mode of communication is different. Books approach the learner through writing, and induce him to write in response. In contrast, MOOCs approach the learner in part through speech, and induce him to speak in response. Speech, for Socrates, is personal and interactive; speech is the road to the soul. However, spoken bilateral interaction cannot occur between a teacher and 20,000 online learners spread over time and space. That format is the ultimate insult to Socratic learning.

Having said that, the networking that can accompany a MOOC may possibly facilitate the internalization of the teacher's message even more effectively than a one-on-one tutorial. Online chats and other modes of networking are fast and multipersonal and can help the learners to rapidly find their own mental contexts for assimilating and modifying the teacher's message. Furthermore, MOOCs can offer the learner alternative pathways depending on progress, ability, and interest. The learning histories of myriad students allow MOOC designers to use Big Data tools for adapting the MOOC to the needs of diverse students.

---

[25] Plato, *Phaedrus*, 278a.

Many people have complained that the Internet undermines the permanence of the written word. No document is final if it's on the Web. Socrates might have approved, and this might be the greatest strength of the MOOC: no course ever ends and no lecture is really final. If MOOCs really are democratic, then they cannot be controlled. The discovery of knowledge, like the stars' movement through space is forever ongoing, with occasional supernovas that brighten the heavens. The creation of knowledge will never end because the unknown is limitless. If MOOCs facilitate this creation, then they are good.

However, the distance and impersonality of MOOCs may impede soul-touching and mind-modifying interactions between teacher and learner. If MOOCs supplant one-on-one interactions between great minds—mature teacher and maturing learner— then their contribution to nurturing an innovative elite is limited at best.

A verdict on MOOCs is hard to reach because MOOCs present innovation dilemmas on two levels. First, the full implications of this new use of information technology and the Internet are still unknown, and will probably continue to harbor surprises as the technology develops. This is the standard innovation dilemma of any new and promising technology.

Second, MOOCs illustrate an innovation dilemma that is inherent in the egalitarian imperative to nurture creative individualism throughout the population. Russell's argument "in praise of idleness" was, in part, that leisure sometimes leads to socially valuable innovation, and that this leisure should be extended to all people for both moral and pragmatic reasons. Morally, because everyone is humanly equal; pragmatically, because you don't know who carries the spark of genius, and social rank is no indication. From the pragmatic perspective, the attempt to widely nurture individualism may in fact impede the education of those unique individuals of special genius. Mass education may tend to be tailored for the bulk of the population and not to the potential elite. In the case of MOOCs this results in distancing the great teacher from the brilliant learner. This innovation dilemma arises from a paradox in the universal principle of egalitarianism as a pragmatic tool for nurturing exceptional individuals. We now explore more deeply this paradox of universalism.

The Paradox of Universalism

A precept or law is universal if it applies everywhere at all times. No exceptions or violations are tolerable or even possible. Universalism can tolerate no deviation. But some universalisms are paradoxical, and some of these paradoxes cause innovation dilemmas.

The foremost example of universalism is the physicist's law of nature. If there are stable universal laws of nature, then no physical event can violate those laws. Any apparent contradiction between observation and law must indicate an error, either in the law or in the observation. For instance, Enrico Fermi studied the apparent non-conservation of energy in beta decay of radioactive elements. This non-conservation contradicts the first law of thermodynamics which states that the total energy of an isolated system is constant, and that energy can be neither created nor destroyed. In 1934 Fermi adopted Pauli's suggestion of the existence of an unobserved sub-atomic particle—which Fermi called the neutrino—that is created during the beta decay event and that carries away the "missing" energy, thus reconciling beta decay with the conservation of energy.[26]

The imperative for survival is also a type of universalism. Biological organisms, industrial corporations, and political states and other human communities all feel an imperative to survive;

[26] A discussion of Fermi's contribution, and a translation of his original paper, is found in Fred L. Wilson, 1968, "Fermi's theory of beta decay", *American Journal of Physics*, 36(12), pp. 1150–60.

no other situation is tolerable, conceivable, or acceptable. Survival is a conservative process of maintaining the entity in a stable state, come what may. This imperative creates what Donald Schön called a "conservative dynamism" in discussing corporate attempts to maintain themselves. However, Schön continues, "it is the crisis of modern industrial corporations that they are also required to undertake technical change destructive of their stable states in order to survive."[27] The paradox of the universal struggle of organisms to survive, at all times and in all conditions, is that the conservatism of survival requires innovation and change. An innovation dilemma arises when on one hand the change promises survival, but on the other hand there is uncertainty about what actually will survive. Without change the organism might vanish, but with change the organism might survive in unrecognizable form.[28]

Military and geopolitical strategy is a type of universalism when it is intended to support tactical or operational decisions in concrete situations. For example, a strategic principle for a small country may be to move the battle to enemy territory as quickly as possible, due to lack of geographical depth. No other plan of battle would be acceptable. However, the strategist knows that strategic principles will sometimes be invalid in practice because of unanticipated contingencies. For instance, a domestic insurgency provides no foreign territory to which the conflict can be moved. The challenge facing the military strategist is to balance between

[27] Donald A. Schön, 1967, *Technology and Change: The New Heraclitus*, Delacorte Press, New York, p. 73.

[28] The paradoxical imperative for survival is reminiscent of the paradox of norms for innovation. Schön writes: "The concept of an ethic of change very nearly appears as a contradiction in terms. Our norms are precisely norms for stability.... Our moral heroes—Socrates excepted—are generally those who stand firm in the face of challenge". (Schön, 1967, p. 204) Norms such as "discovery" or "boldness in exploration" (p. 214) underlie innovation and change, but they must be immunized against alteration in order to protect innovation.

fundamental long-range strategic principles, and pragmatic solution of pressing problems. At one extreme the strategist ignores tactical contingencies and insists on adherence to strategic principles. At the other extreme, the strategist abdicates and devotes all innovation and initiative to the solution of specific problems.

For instance, Raphael D. Marcus explains that the Israel Defense Forces traditionally tended to focus on developing specific solutions to particular problems. In "Israel's early military history, [this] sometimes led to…a general dearth of deep strategic thinking. However,…constant tactical innovation can be considered a major strength of the IDF".[29] Similarly, James A. Russell writes that US brigade and battalion commanders fighting insurgencies in Iraq from 2005 to 2007 were "Unconstrained by an established [counterinsurgency] doctrine,…[had] wide flexibility to structure their operations [and] freely developed a series of new organizational capacities".[30] In both of these examples, the absence of prevailing strategic doctrine facilitated innovative bottom-up solutions to tactical and operational problems. However, the lack of strategy can have adverse long-term implications due to lack of system-wide foresight and inadequate force development and training.

The military strategist's challenge—balancing principles against pragmatism—is a paradox of universalism. A strategic principle is universal if it applies everywhere at all times. No exceptions or violations are tolerable. The paradox of universalism is that unknown future contingencies may force operational violation of the principle.

[29] Raphael D. Marcus, 2015, "Military innovation and tactical adaptation in the Israel–Hizballah Conflict: The institutionalization of lesson-learning in the IDF", *Journal of Strategic Studies*, 38(4), pp. 500–28; p. 508.

[30] James A. Russell, 2010, "Innovation in war: Counterinsurgency operations in Anbar and Ninewa Provinces, Iraq, 2005–2007", *Journal of Strategic Studies*, 33(4): pp. 595–624, p. 597.

The concept of an innovation dilemma assists in understanding the strategist's challenge in any field, military or civilian. An innovative and highly promising new strategy is less familiar than a more standard strategic approach whose tactical implications are more familiar. Hence the innovation, while purportedly better than the standard approach, may be much worse due to uncertainty about the innovation. The resolution (never unambiguous) of the dilemma results from analysis of robustness to surprise, as discussed in Chapter 5.

Political and moral precepts are often universalistic, and sometimes entail painful paradoxes. For example, consider the historically recurring tension between the principle of freedom and the principle of equality, which is one of the oldest quandaries in political philosophy, as explained by C. Northcote Parkinson.[31]

Freedom. People have widely varying interests and aptitudes, and some people are far more talented than others, as we discussed earlier (see p. 117). A society that offers broad freedom for individuals to exploit their abilities will also develop a wide spread of wealth, accomplishment, and status. Freedom, when supported by adequate leisure time, enables unique individuals to explore, invent, discover, create, and acquire both intellectual and material wealth. Freedom and leisure are the recipe for innovation, as Bertrand Russell explained (see p. 118). Freedom creates inequality because talents and propensities vary among people.

Equality. People vary widely in their interests, aptitudes, and abilities. Hence equality among members of a society— meaning similar accomplishment, status, and wealth, and not just equality of opportunity—is not an inherent or unavoidable social condition. A society that strives for equality among its members can achieve this by enforcing conformity and by

[31] C. Northcote Parkinson, 1958, *The Evolution of Political Thought*, University of London Press, London.

transferring material wealth from rich to poor. In democratic societies, wealth is usually transferred coercively through obligatory taxation. Conformity is often obtained by social mores and opprobrium for deviation, but without legal or physical coercion. One benefit of social equality is that it provides a degree of security, a personal and social safety net. The implementation of equality by coercion and opprobrium may, however, tend to reduce the incentive for unique individuals to exploit their freedom for innovation.

The dilemma is that a life without the freedom to invent and discover is hardly human, but freedom without security is the jungle. And life in the jungle, as Hobbes explained, is "solitary, poor, nasty, brutish and short".[32]

Let's consider a different example of the paradox of universalism in political and moral precepts. Following the horrors of the Second World War, the United Nations adopted the Universal Declaration of Human Rights (UDHR) in "recognition of the inherent dignity and of the equal and inalienable rights of all members of the human family" and in aspiring to promote "freedom, justice and peace in the world". The precepts expounded in the UDHR are not as precise as the quantitative laws of physics. They nonetheless are "a common standard of achievement for all peoples and all nations, to the end that every individual and every organ of society...shall strive...to secure their universal and effective recognition and observance".[33]

The universalism of this admirable document entails paradoxes, and these reveal an underlying innovation dilemma. Consider, for instance, Article 19, which asserts:

---

[32] Thomas Hobbes, 1668, *Leviathan: or the Matter, Form and Power of a Commonwealth Ecclesiastical and Civil*, chapter XIII.

[33] See http://www.un.org/en/documents/udhr, accessed November 11, 2014.

Everyone has the right to freedom of opinion and expression; this right includes freedom to hold opinions without interference and to seek, receive and impart information and ideas through any media and regardless of frontiers.

The paradox is that toleration for expressing and transmitting all opinions includes toleration for incitement against toleration. Examples include incitement to racism, or sexism, or terrorism, or other actions that subvert a tolerant pluralistic social order. If pluralistic toleration of diversity of opinion is meant to be sustainable, then the dissemination of attitudes that subvert pluralistic toleration must be restricted. All liberal democratic States limit expression that entails clear and present danger of subverting recognized freedoms (such as freedom of expression), while allowing all other expression. So, while the Declaration is universal, its application is restricted to some extent by all States that adhere to it.

Article 19 is paradoxical in another sense as well, even after we modify it to protect tolerant pluralism. If everyone agrees to a precept, for example, that pedophilia is intolerably abhorrent, then prohibiting the promotion of pedophilia falls in the category of legitimately prohibited communication. But suppose (as is the case) that humanity is divided on a fundamental precept, for example that insulting a holy and revered entity is intolerably abhorrent. If irreverent speech of this sort is allowed, then "inherent dignity" has been denied to people, whose most cherished values are trampled.[34] However, if irreverent speech is disallowed, then other people are hurt in the same way. In other

[34] For example, J. David Goodman wrote in the *New York Times* (December 29, 2010) that "satirical cartoons of Muhammad were commissioned and published [on September 30, 2005] by [the Danish newspaper] Jyllands-Posten as a statement of freedom of expression. But they were seen as blasphemous and a deliberate provocation by many Muslims, and prompted rioting in some countries and repeated attempts at violent retribution."

words, Article 19 seems to assume that humanity has pretty much reached consensus on fundamental issues of truth, morality, religion, and so forth. But the framers of the UDHR assumed nothing of the sort. They acknowledged the need to "strive by teaching and education to promote respect for these rights and freedoms and by progressive measures, national and international, to secure their universal and effective recognition and observance", as stated in the Preamble. The paradox—contradiction—is the presumption that "education to promote respect for these rights and freedoms" can be done without trampling values that are most cherished by some segment of humanity. The framers clearly recognized the uphill struggle to extirpate mores and beliefs that are contrary to the ideals of the UDHR. The paradox embodied in the Declaration is that this extirpation will trample the inherent dignity of people whose mores and beliefs are unacceptable to the framers.

The UDHR also illustrates that the paradox of universalism can induce an innovation dilemma. Uncertainty is the source of the innovation dilemma, so we will expand on the uncertainties inherent in the UDHR.

The UDHR was adopted by the UN General Assembly on December 10, 1948. The task of formulating the document was tremendously difficult because the world was divided into Eastern and Western blocs. Eleanor Roosevelt, who chaired the drafting committee, noted in her memoirs an exchange between Charles Malik from Lebanon and Peng Chung Chang of China. Dr Malik expounded at length precepts from Thomas Aquinas, while Dr Chang countered by urging the in-depth study of Confucian philosophy.[35] The drafters of the UDHR had to bridge substantial cultural gaps in order to reach consensus. Yet nobody could have

---

[35] See http://www.un.org/en/documents/udhr/history.shtml, accessed November 11, 2014.

anticipated other philosophies or ideological movements, such as radical Islam, that would come to the forefront of human society in the future. The future cultural milieu in which the Declaration would reside could not have been known to the drafting committee as it toiled away in 1947 and 1948.

A different uncertainty is contained in Article 19. While clearly intended to support a universal value of great importance, the article entails great uncertainty. The last phrase protects communication "through any media and regardless of frontiers." "Any media" refers to media not invented when the Declaration was drafted, such as the Internet or social media, whose potency could not be anticipated and yet is allowed. The power of social media to topple dictators was illustrated dramatically in Tunisia, Egypt, Libya, Yemen, and other countries during 2011. And yet, it remains to be seen whether or not these transformations will eventually lead to greater adherence to the principles of the UDHR.[36] The promise of free exchange of ideas and information "regardless of frontiers" protects the terrorist sending messages of hate and incitement (or instructions for bomb assembly) across international borders and around the globe at the touch of a keypad, just as much as it protects the searing cries of poets and prophets.

An innovation dilemma arises from the preference for a putatively better but more uncertain alternative. The committee that drafted the UDHR toiled for two years, and many different documents could have resulted. However, when the final draft was brought before the UN General Assembly, member States had two options: accept or reject the draft.[37]

---

[36] At the time of writing we are seven years after the initial upheavals.

[37] For simplicity, we will ignore the possibility of returning the draft to the committee.

The Universal Declaration of Human Rights was a daring and important innovation on the international scene in 1948. The Declaration embodies fundamental moral and social values that were supported by representatives from a broad spectrum of cultures. Its universalism strengthens its appeal. In the immediate aftermath of the Second World War, the Declaration seemed invaluable in rebuilding the world as a better place. But the uncertainty of its future interpretation and application is the source of the dilemma. The innovation dilemma arises because one cannot know if the cultural values and attitudes that are protected will undermine those very values and even strengthen the enemies of those values. To refrain from supporting the Declaration leaves the field open for cultural regression and conflict. To adopt the Declaration runs the risk of protecting values (unidentified at the time) that are abhorrent to the drafters of the Declaration. The decision to adopt or not adopt the UDHR entails a dilemma resulting from its putative value and the vast uncertainties surrounding its future impact.

We have seen that a paradox of universalism can induce an innovation dilemma. One way to resolve the dilemma is to deny the universality and thus to remove the paradox. For example, if free expression is not universal, but limited only to particular expressions (those of "our" culture but not of "theirs"), then the paradox of universal free speech is resolved and the dilemma is ameliorated. However, this resolution of the innovation dilemma introduces the risk of relativism: truths are not universal, but only relative to a particular culture or world view. Some moral stands are indeed culturally linked, such as the acceptance of polygamy in many ancient civilizations. Having said that, for some propositions, such as the Universal Declaration of Human Rights or the physicist's laws of nature, the universalism cannot be denied without fundamentally altering the meaning or the intention of the propositions.

The search for truths—universally valid propositions—is a hallmark of innovative societies. Non-innovative societies already have their full complement of truths; innovative societies are continually on the lookout for new universalisms and old mistakes. The dilemmas entailed in the paradox of universalism are therefore an inevitable correlate of innovation. While innovations can strengthen and improve a society, they can also engender fragility. This is the innovation dilemma of universalism.

The Fragility of Innovative Societies

Innovative societies may be fragile in two senses. First, innovation may tend to undermine the shared values that facilitate innovation in the first place. We will refer to this as *value-based fragility.* Second, innovation may tend to undermine the institutional basis for cooperation that is necessary for innovation to take place. We'll call this *institutional fragility.* Neither of these mechanisms is rigidly deterministic or unavoidable; understanding the potential fragilities can help in avoiding adverse outcomes and in managing the innovation dilemmas that are inherent in innovation.

We begin by discussing **value-based fragility.** The outline of our argument is as follows.

1. Innovation is facilitated by cooperation.
2. Cooperation is facilitated by shared values.
3. Shared values may be eroded by innovation.
4. Thus innovation may erode cooperation and inhibit innovation.

Teamwork is an essential element of much innovation. Consider for example innovative engineering design. Much technological innovation is an incremental step-by-step process whose long-term impact is unknown and may be revolutionary. An initial brainstorming stage combines the insights and intuitions of professionals in various disciplines as well as stakeholders and other interested parties. The next stage is the formulation of basic design concepts based on discussion and debate among technical specialists from relevant disciplines as well as end users and other participants. The final detailed design will be the coordinated effort of technical specialists in various fields. In Chapter 5 we discussed an example of this sort of co-design for bringing water from a river into a small rural town in Central America.

Other examples of cooperation in innovation arise in international affairs, for instance the United Nations' adoption of the Universal Declaration of Human Rights discussed earlier. This admirable and innovative document was the fruit of collaboration between scholars and leaders from diverse cultural backgrounds. Its claim to universality depended intimately on the collaborative nature of the effort.

Shared values are the basis of much collaborative effort, for which the UN Declaration is a prime example. Indeed, the heart of that effort was to identify human values that are shared universally. There is no need to deny that self interest is also often a legitimate basis for cooperation. Adam Smith famously wrote that "It is not from the benevolence of the butcher, the brewer, or the baker, that we expect our dinner, but from their regard to their own interest."[38] Nonetheless, sharing social or moral values provides additional motivation for cooperation. For instance, Smith's small-town food producers may very well have valued the sense of community instilled by their daily business routine.

Innovation can erode shared values. The history of technology is rife with examples of inventions altering social values. For instance, as we discussed on p. 112, Landes explains that early mechanical clocks summoned the faithful to prayer in medieval monasteries.

[38] Adam Smith, 1776, *An Inquiry into the Nature and Causes of the Wealth of Nations*, Book I, Chapter II, Prometheus Books, Amherst, NY, third paragraph.

But the long-term influence of the mechanical clock on Western civilization was the idea of

> *time discipline* as opposed to *time obedience*. One can ... use public clocks to summon people for one purpose or another; but that is not punctuality. Punctuality comes from within, not from without. It is the mechanical clock that made possible, for better or for worse, a civilization attentive to the passage of time, hence to productivity and performance.[39]

One may view time discipline as either an erosion or an improvement in social values. In either case the example demonstrates how a technological innovation altered a social value.

Examples abound of innovations altering values. The development of physical science and its naturalistic understanding of "celestial mechanics"[40] in seventeenth- and eighteenth-century Europe contributed to the decline of theological attitudes to "the heavens".[41] The Internet and its impact on interpersonal relations is another case in point. Social media enable one to communicate extensively and intimately with people you've never met. This globally extends the scope of one's acquaintances, but it can also diminish one's connection with the immediate surroundings and reduce the value one ascribes to them. Medically efficient contraceptives have contributed to modifying attitudes to extramarital sex and the social institution of the family.

Finally, *innovation may erode the willingness or ability to cooperate*. As values change, the propensity to cooperate in traditional frameworks can change as well. Smith's provincial butcher, brewer, or baker may find that globalization alters their market

[39] Landes, *Revolution in Time*, p. 7 (emphasis in original).

[40] Pierre-Simon, Marquis de Laplace published his five-volume treatise *Celestial Mechanics* from 1799 to 1825, in which he formalized and extended earlier work by Newton and many others.

[41] Arthur Koestler, 1959, *The Sleepwalkers: A History of Man's Changing Vision of the Universe*, Hutchinson, London.

opportunities as well as their community self-identity. What was traditionally a second-nature preference for the local "us" may become a globalized identity of "we all". The same goes for their customers who have no compunction in patronizing huge chain franchises at the expense of the neighborhood store. This is not only the pocketbook speaking. It is the pocketbook expressing different social values that emerged as a result of innovations in production and marketing.

Most of the innovations we have considered so far in this discussion of the value-based fragility of innovative societies have been technological. However, innovations can also be entirely conceptual. For example, social organization—including cooperation among members of a society—is intimately linked to ideas and values. As a case in point, a major ideological innovation was the emergence of moral and political conceptions of the equality and dignity of Man (or at least of men) in eighteenth-century Western societies. This new spirit of equality undermined the social organization of monarchical societies and eroded the "cooperation" between aristocracies and other classes of society. A subsequent wave of this conceptual innovation occurred during the early twentieth century when women were explicitly included in the family of Man, and were granted the vote, encouraged to work outside of the home, and so on. This led to various changes in social relations—cooperative and others—between the sexes.

In summary, it has been explained that innovation often depends on cooperation that is supported by shared values that can be eroded by innovations. This mechanism for "negative feedback" of innovation on itself is what we are calling valued-based fragility of innovative societies. This mechanism is not inevitable or unalterable. We are not making a Spengler-like prediction of cyclical ups and downs of innovative societies.[42]

[42] Oswald Spengler, *Decline of the West.*

We have explained that innovations can adversely influence the process of innovation. This creates an innovation dilemma, especially for a society that views innovation as a value in itself. We innovate because that is valuable in and of itself, but some innovations may tend to erode the ability to innovate. We don't know beforehand whether a given innovation will negatively influence the overall innovative process; hence the dilemma.

This paradox also arises through the mechanism that we are calling **institutional fragility** of innovative societies. The rough outline of this mechanism is:

1. Innovation is facilitated by cooperation.
2. Cooperation facilitates specialization.
3. Specialization strengthens competition.
4. Competition may undermine cooperation and inhibit innovation.

Institutional and value-based fragility are interrelated. Indeed, the first step is the same in both arguments: innovation is facilitated by cooperation. But then institutional fragility diverges to emphasize organizational aspects.

Specialization of labor depends on cooperation among specialists. Robinson Crusoe had no one with whom to cooperate, so he could not specialize in any of the functions essential for his survival. He learned, and improved, but could never reach full mastery of any specific activity, and many activities (violin making, for instance) he never attempted at all. Adam Smith's butcher, however, cooperates with the brewer by exchanging meat for beer. In early human hunter-gatherer bands the stone-tool maker cooperated with the hunter for the benefit of both. Today the neurosurgeon can cooperate with the stockbroker (and usually at least one of them benefits).

Specialization enhances competitive advantage. The competitive advantage of specialists derives in part from the higher quality, or efficiency, or productivity of their work. From an organizational point of view, specialization creates a holistic advantage that is greater than the sum of the parts. For instance, Adam Smith explains that one man can produce at most twenty pins in a day's work, while ten specialists can cooperatively produce "upwards of forty-eight thousand pins in a day. Each person, therefore,... might be considered as making four thousand eight hundred pins in a day."[43]

Competition tends to undermine cooperation among specialists of the same field. Specialists in all fields, whether pin production or neurosurgery, will often share their innovations for the benefit of all.[44] This results, in part, from shared values (e.g. the ethos of a scientific or professional community), as well as shared interests (you help me and I'll help you). Nonetheless, there is also a strong propensity to guard one's specialist advantage against competitors.

The patent system is one mechanism by which competition may undermine cooperation. Boldrin and Levine write that "Property is a good thing. Ownership...provides incentive to produce, accumulate, and trade."[45] Patents and copyrights are one form of ownership, and they have many important benefits both for inventors and for society at large. However, patents and copyrights can also inhibit cooperation and thus impede innovation. Boldrin and Levine discuss the case of James Watt, who was

> one of many clever inventors who worked to improve steam power in the second half of the eighteenth century. After getting one step ahead of the pack, he remained ahead not by superior

[43] Smith, *Wealth of Nations*, Book I, chapter I, third paragraph.

[44] Michele Boldrin and David K. Levine, 2008, *Against Intellectual Monopoly*, Cambridge University Press, Cambridge, p. 136.

[45] Boldrin and Levine, *Against Intellectual Monopoly*, p. 123.

innovation but also by superior exploitation of the legal system. The fact that his business partner was a wealthy man with strong connections in Parliament was not a minor help....

In the specific case of Watt, the granting of the 1769 and especially of the 1775 patents likely delayed the mass adoption of the steam engine: innovation was stifled until his patents expired, and few steam engines were built during the period of Watt's legal monopoly.[46]

In summary, the institutional fragility of innovative societies derives from the competitive specialization that facilitates many innovations. Specialization depends on cooperation, but competition may inhibit cooperation. The net effect can be a self-limitation of innovation. Like value-based fragility, this institutional fragility is not deterministic or inevitable. Both mechanisms are potential impediments of innovation that, once identified, can be grappled with and managed to some degree. The fragility of innovative societies is an innovation dilemma in its most generic realization: the process of innovation may potentially undermine the continued process of innovation, and this presents the innovative society with a dilemma. The personal or social drive to innovate is the source of progress, but innovation is uncertain, and the drive to optimize the human condition may not be optimal.

[46] Boldrin and Levine, *Against Intellectual Monopoly*, pp. 2–3.

References

Adams, Henry, 1918, *The Education of Henry Adams*, edited by Ernest Samuels, Houghton Mifflin Co., Boston.

Aristotle, *Politics*, in McKeon, ed., 1941.

Bell, David E., Howard Raiffa, and Amos Tversky, eds., 1988, *Decision Making: Descriptive, Normative, and Prescriptive Interactions*, Cambridge University Press, Cambridge, UK.

Bellman, Richard, 1961, *Adaptive Control Processes: A Guided Tour*, Princeton University Press, Princeton, NJ.

Ben-Haim, Yakov, 2006, *Info-gap Decision Theory: Decisions Under Severe Uncertainty*, 2nd edn., Academic Press, London.

Ben-Haim, Yakov, 2007, "Peirce, Haack and info-gaps", in *Susan Haack, A Lady of Distinctions: The Philosopher Responds to Her Critics*, Cornelis de Waal, ed., Prometheus Books, Amherst, NY.

Ben-Haim, Yakov, 2010, *Info-Gap Economics: An Operational Introduction*, Palgrave Macmillan, London.

Ben-Haim, Yakov, 2012, "Doing our best: Optimization and the management of risk", *Risk Analysis*, 32(8), pp. 1326–32.

Ben-Haim, Yakov, 2015, "Dealing with uncertainty in strategic decision-making", *Parameters*, US Army War College Quarterly, 45(3), Autumn 2015.

Ben-Haim, Yakov, Craig D. Osteen, and L. Joe Moffitt, 2013, "Policy dilemma of innovation: An info-gap approach", *Ecological Economics*, vol. 85, pp. 130–8.

Bentham, Jeremy, 1776, *A Fragment on Government*, in Harrison, ed., 2000.

Boldrin, Michele and David K. Levine, 2008, *Against Intellectual Monopoly*, Cambridge University Press.

Boyne, Walter J., 2008, "Goering's big bungle", *Air Force Magazine*, 91(11), November 2008, pp. 58–61.

Burns, Robert, 1785, "To A Mouse", *The Kilmarnock Edition of the Poetical Works of Robert Burns*, William Scott Douglas, ed., the Scottish *Daily Express*, Glasgow, 1938.

Burt, Vivien, Caryn Bernstein, Wendy S. Rosenstein, and Lori L. Altshuler, 2010, "Bipolar disorder and pregnancy: Maintaining psychiatric stability in the real world of obstetric and psychiatric complications", *American Journal of Psychiatry*, 167(8), pp. 892–7.

Burtt, Edwin A., ed., 1939, *The English Philosophers From Bacon to Mill*, The Modern Library, New York.

Canadian Security Intelligence Service, 2016, *Al-Qaeda, ISIL and Their Offspring*, Highlights from the workshop, February 29, 2016, CSIS National Headquarters, Ottawa, Canada.

Carey, J. R., F. G. Zalom, and B. D. Hammock, 2008, Concerns with the eradication program against the light brown apple moth in California. Personal communication to E. Schafer on May 28.

Carmel, Yohay and Yakov Ben-Haim, 2005, "Info-gap robust-satisficing model of foraging behavior: Do foragers optimize or satisfice?" *The American Naturalist*, vol. 166, pp. 633–41.

Carroll, Lewis (Charles Lutwidge Dodgson), 1865, *Alice's Adventures in Wonderland*, Project Gutenberg, Salt Lake City, UT, last updated July 14, 2014.

Carroll, Lewis (Charles Lutwidge Dodgson), 1871, *Through the Looking-Glass*, Project Gutenberg, Salt Lake City, UT, last updated October 20, 2015.

Chen, Ingfei, 2010, "From medfly to moth: Raising the buzz of dissent", *Science*, 327(5962), pp. 134–6, January 8, 2010.

Christensen, Clayton M., 2011, *The Innovator's Dilemma: The Revolutionary Book That Will Change the Way You Do Business*, Harper Business, New York.

Clagett, Marshall, 1955, *Greek Science in Antiquity*, Collier Books, New York.

Dempster, A. P., 1967, "Upper and lower probabilities induced by multi-value mappings". *Annals of Mathematical Statistics*, vol. 38, pp. 325–39.

Dirac, P. A. M., 1958, *The Principles of Quantum Mechanics*, 4th edn., Oxford University Press, Oxford.

Edwards, P., ed., *The Encyclopedia of Philosophy*, Simon and Schuster Macmillan, vol. 5.

Einstein, A., 1905, "Zur Elektrodynamik bewegter Körper" ("On the Electrodynamics of Moving Bodies"), *Annalen der Physik*, (ser. 4), vol. 17, pp. 891–921.

Engineers without Borders, http://www.ewb-uk.org, accessed June 26, 2016.

Evans, George W. and Seppo Honkapohja, 2001, *Learning and Expectations in Macroeconomics*, Princeton University Press, Princeton, NJ.

Ferdman, Roberto A., 2015, "Bye bye, bananas", *Washington Post*, December 4, 2015.

Ferson, S., 2002, *RAMAS Risk Calc 4.0 Software: Risk Assessment with Uncertain Numbers*, Lewis Publishers, Boca Raton, FL.

Feynman, Richard P., 1948, "Space-time approach to non-relativistic quantum mechanics", *Reviews of Modern Physics*, 20(2), pp. 367–87.

Fingar, Thomas, 2011, *Reducing Uncertainty: Intelligence Analysis and National Security*, Stanford Security Studies, Stanford, CA.

Fischer, R. A., Derek Byerlee, and G. O. Edmeades, 2009, "Can Technology Deliver on the Yield Challenge to 2050?" Expert Meeting on How to feed the World in 2050, Economic and Social Development Department, Food and Agriculture Organization of the United Nations, Rome, June 24–6, 2009.

Forbes online, "5 Best Ways to Maximize Your Retirement Investments", March 16, 2012, http://www.forbes.com/sites/thestreet/2012/03/16/5-best-ways-to-maximize-your-retirement-investments accessed October 22, 2015.

Franklin, James, 2001, *The Science of Conjecture: Evidence and Probability before Pascal*, The Johns Hopkins University Press, Baltimore, MD.

Galbraith, John Kenneth, 1986, *The New Industrial State*, 4th edn., Mentor Books, Penguin Group, New York.

Goodman, J. David, "Police Arrest 5 in Danish Terror Plot", *New York Times*, December 29, 2010.

Hacking, Ian, 1975, *The Emergence of Probability: A Philosophical Study of Early Ideas About Probability, Induction and Statistical Inference*, Cambridge University Press, Cambridge, UK.

Hamilton, Edith and Huntington Cairns, eds., 1961, *The Collected Dialogues of Plato, including the Letters*, Princeton University Press, Princeton, NJ.

Harder, Daniel, and Jeff Rosendale, 2008, "Integrated Pest Management Practices for the Light Brown Apple Moth in New Zealand: Implications for California", March 6, 2008, 14 pp, http://www.lbamspray.com/00_Documents/2008/HarderNZReportFINAL.pdf, accessed July 24, 2011.

Harder, Daniel, Ken Kimes, Roy Upton, and Lynette Casper, 2009, "Light Brown Apple Moth (LBAM) Eradication Program: Formal Petition to Recalssify LBAM as a Non-quarantinable Pest: Summary of Findings", January 7, 2009, http://www.lbamspray.com/Reports/Reclassification Petition SummaryRU.pdf, accessed July 3, 2011.

Harkabi, Yehoshafat, 1990, *War and Strategy*, (in Hebrew), Maarachot Publishers, Israel Ministry of Defense, Tel Aviv.

Harrison, Ross, ed., 2000, *Selected Writings on Utilitarianism: Bentham*, Wordsworth Classics, Ware, Hertfordshire, UK.

Heath, Thomas, 1921, *A History of Greek Mathematics*: vol. 1: *From Thales to Euclid*; vol. 2: *From Aristarchus to Diophantus*, Clarendon Press, Oxford, reissued by Dover Publications, New York, 1981.

Heath, Thomas, ed., 1925, *The Thirteen Books of Euclid's Elements*, 2nd edn., Cambridge University Press, reissued by Dover Publications, New York, 1956, vol. 1.

Heaton, E. W., 1974, *Solomon's New Men: The Emergence of Ancient Israel as a National State*, Thames & Hudson, London.

Hobbes, Thomas, 1668, *Leviathan: or the Matter, Form and Power of a Commonwealth Ecclesiastical and Civil*, in Burtt, 1939.

Hume, David, 1748, *An Enquiry Concerning Human Understanding*, in Burtt, 1939.

Huntington, Samuel P., 1957, *The Soldier and the State: The Theory and Politics of Civil–Military Relations*, Harvard University Press, Cambridge, MA.

Kahn, Herman, 1961, *On Thermonuclear War*, 2nd edn., Princeton University Press, Princeton, NJ.

Keegan, John, 1994, *A History of Warfare*, Pimlico, London.

Keynes, John Maynard, 1920, *The Economic Consequences of the Peace*, Harcourt, Brace and Howe, New York.

Keynes, John Maynard, 1936, *The General Theory of Employment, Interest, and Money*, Harcourt Brace & World, reissued by Prometheus Books, Amherst, NY, 1997.

Kilcullen, David, 2013, *Out of the Mountains: The Coming Age of the Urban Guerrilla*, Hurst & Company, London.

Klein, Jacob, 1992, *Greek Mathematical Thought and the Origin of Algebra*, reissued by Dover Publications, New York.

Klein, Jacob, "Aristotle, an introduction", in Williamson and Zuckerman, eds., 2013, pp. 171–95.

Kline, Morris, 1972, *Mathematical Thought From Ancient to Modern Times*, 3 vols., Oxford University Press, Oxford.

Klir, G. J., 2006, *Uncertainty and Information: Foundations of Generalized Information Theory*, Wiley, New York.

Knight, Frank H., 1921, *Risk, Uncertainty and Profit*, Hart, Schaffner and Marx, reissued by Harper Torchbooks, New York, 1965.

Knight, Frank H., 1933, *The Economic Organization*, reissued by Harper Torchbooks, New York, 1951.

Koestler, Arthur, 1959, *The Sleepwalkers: A History of Man's Changing Vision of the Universe*, Hutchinson, London.

Koestler, Arthur, 1967, *The Ghost in the Machine*, Hutchinson, London.

Krepinevich, Andrew F. and Barry D. Watts, 2015, *The Last Warrior: Andrew Marshall and the Shaping of Modern American Defense Strategy*, Basic Books, New York.

Kuhn, Thomas S., 1970, *The Structure of Scientific Revolutions*, 2nd edn., University of Chicago Press, Chicago.

Lanczos, Cornelius, 1970, *The Variational Principles of Mechanics*, 4th edn., Dover Publications, New York.

Landes, David S., 1983, *Revolution in Time: Clocks and the Making of the Modern World*, Harvard University Press, Cambridge, MA.

Laplace, Marquis de (Pierre-Simon) *A Philosophical Essay on Probabilities*, trans. by F. W. Truscott and F. L. Emory, 1951, Dover Publications, New York.

Lejewski C., 1996, "Jan Lukasiewicz", in P. Edwards, ed., *The Encyclopedia of Philosophy*, vol. 5, Simon and Schuster Macmillan, New York.

Lewis, John David, 2010, *Nothing Less than Victory: Decisive Wars and the Lessons of History*, Princeton University Press, Princeton, NJ.

Lindgren, Astrid, 1945, *Pippi Longstocking*, Viking Press, New York.

Locke, John, 1706, *An Essay Concerning Human Understanding*, 5th edn., Roger Woolhouse, ed., Penguin Books, London, 1997.

Luenberger, David G., 1969, *Optimization by Vector Space Methods*, Wiley, New York.

March, James G. 1988, "Bounded rationality, ambiguity, and the engineering of choice", in Bell, Raiffa, and Tversky, eds., 1988.

Marcus, Raphael D., 2015, "Military innovation and tactical adaptation in the Israel–Hizballah Conflict: The institutionalization of lesson-learning in the IDF", *Journal of Strategic Studies*, 38(4), pp. 500–28.

Mayo, Deborah G., 1996, *Error and the Growth of Experimental Knowledge*, University of Chicago Press, Chicago.

McKeon, Richard, ed., 1941, *The Basic Works of Aristotle*, Random House, New York.

Mill, John Stuart, 1861, *Utilitarianism*, in Burtt, ed., 1939.

Mill, John Stuart, 1869, *The Subjection of Women*, Longmans, Green, Reader & Dyer, London.

Miranda, E., 2008, "A survey of the theory of coherent lower previsions", *International Journal of Approximate Reasoning*, vol. 48, pp. 628–58.

Overy, Richard, 2006, *Why the Allies Won*, 2nd edn., Pimlico, London.

Parkinson, C. Northcote, 1958, *The Evolution of Political Thought*, University of Press, London.

Oxford English Dictionary, www.oed.com.

Plato, *Phaedrus*, in Hamilton and Cairns, eds., 1961.

Plato, *Republic*, in Hamilton and Cairns, eds., 1961.

Plato, *Timeaus* and *Critias*, in Hamilton and Cairns, eds., 1961.

Popper, Karl R., 1961, *The Poverty of Historicism*, reissued by Harper Torchbooks, New York.

Popper, Karl R., 1968, *The Logic of Scientific Discovery*, Harper & Row, New York.

Popper, Karl R., 1982, *The Open Universe: An Argument for Indeterminism.* From the Postscript to *The Logic of Scientific Discovery*, reissued by Routledge, London and New York.

Porter, Theodore M., 1986, *The Rise of Statistical Thinking: 1820–1900*, Princeton University Press, Princeton, NJ.

Quenqua, Douglas, 2014, "Is e-reading to your toddler story time, or simply screen time?" October 11, 2014, *New York Times.* http://nyti.ms/1EIAhk6.

Russell, Bertrand, 1932, "In Praise of Idleness", in Bertrand Russell, *In Praise of Idleness and Other Essays*, Unwin Paperbacks, London, 1960.

Russell, Bertrand, 1937, *A Critical Exposition of the Philosophy of Leibniz*, 2nd edn., George Allen and Unwin, London.

Russell, James A., 2010, "Innovation in war: Counterinsurgency operations in Anbar and Ninewa Provinces, Iraq, 2005–2007", *Journal of Strategic Studies*, 33(4), pp. 595–624, p. 597.

Russo, Lucio, 2003, *The Forgotten Revolution: How Science was Born in 300 B.C. and Why It Had to Be Reborn*, with the collaboration of the translator, Silvio Levy, Springer, Berlin.

Savage, Leonard J., 1972, *The Foundations of Statistics*, 2nd revised edn., Dover Publications, New York.

Sayers, D. L., 1928, *Lord Peter Views the Body*, Amereon Ltd.

Schön, Donald A., 1967, *Technology and Change: The New Heraclitus*, Delacorte Press, New York.

Schwartz, Barry, 2004, *Paradox of Choice: Why More Is Less*, Harper Perennial, New York.

Science for Environment Policy, 2013, "Choosing between established and innovative policy measures: controlling invasive species", European Commission DG, Environment News Alert Service, edited by SCU,

The University of the West of England, Bristol, http://ec.europa. eu/environment/integration/research/newsalert/pdf/325na2_en.pdf, accessed March 5, 2018.

Shackle, G. L. S., 1972, *Epistemics and Economics: A Critique of Economic Doctrines*, Cambridge University Press, reissued by Transaction Publishers, New Brunswick and London, 1992.

Shafer, G., 1976, *A Mathematical Theory of Evidence*, Princeton University Press, Princeton, NJ.

Shakespeare, William, 1599, *Henry IV*, Part 2.

Shakespeare, William, *c.*1611, *Cymbeline*.

Simon, Herbert Alexander, 1956, "Rational choice and the structure of the environment", *Psychological Review*, 6(2), pp. 129–38.

Simon, Herbert Alexander, 1997, *Models of Bounded Rationality*, vol. 3, *Empirically Grounded Economic Reason*, MIT Press, Cambridge MA.

Smith, Adam, 1776, *An Inquiry into the Nature and Causes of the Wealth of Nations*, Prometheus Books, Amherst, NY, 1991.

Smolin, Lee, 2006, *The Trouble with Physics: The Rise of String Theory, the Fall of a Science and What Comes Next*, Penguin Books, London.

Snow, C. P., 1966, *Variety of Men*, Charles Scribner's Sons, New York.

Spengler, Oswald, *Decline of the West*, vol. 1: *Form and Actuality*, 1926, vol. 2: *Perspectives of World-History*, 1928, Alfred A. Knopf, New York.

Stigler, Stephen M., 1986, *The History of Statistics: The Measurement of Uncertainty before 1900*, Harvard University Press, Cambridge, MA.

Sun Tzu, *The Art of War*, trans. Samuel B. Griffith, 1963, Oxford University Press, Oxford.

Twain, Mark (Samuel Langhorne Clemens), 1872, *Roughing It*, University of California Press, Berkeley, CA.

Twain, Mark (Samuel Langhorne Clemens), 1884, *The Adventures of Huckleberry Finn*, Harper & Brothers, New York and London.

United States Department of Agriculture, 2008, USDA Budget Explanatory Notes for the Committee on Agriculture for FY 2009, FY 2010, and FY 2011.

United States Department of Agriculture, 2010, Animal and Plant Health Inspection Service Draft Response to Petitions for the Reclassification of Light Brown Apple Moth [Epiphyas postvittana (Walker)] as a Non-Quarantine Pest, Revision January 14, 2010, http://www.aphis. usda.gov/plant_health/plant_pest_info/lba_moth/downloads/draft_ lbam_petition_response-10.pdf, accessed July 2, 2011.

United States Department of Agriculture, 2011, Light Brown Apple Moth, http://www.aphis.usda.gov/plant_health/plant_pest_info/lba_moth/background.shtml, accessed July 2, 2011.

van den Bosch, F., M. J. Jeger, and C. A. Gilligan, 2006, "Disease control and its selection for damaging plant virus strains in vegetatively propagated staple food crops; a theoretical assessment", *Proceedings of the Royal Society, B, Biological Sciences*, vol. 274, pp. 11–18.

Wald, A., 1945, "Statistical decision functions which minimize the maximum risk", *Annals of Mathematics*, 46(2), pp. 265–80.

Walley, P., 1996, "Measures of uncertainty in expert systems", *Artificial Intelligence*, vol. 83, pp. 1–58.

Whitehead, Alfred North, 1925, *Science and the Modern World*, Lowell Lectures, reissued 1948 by Mentor Books, New York.

Williamson, Robert B. and Elliot Zuckerman, eds., 2013, *Jacob Klein Lectures and Essays*, St. John's College Press, Winnipeg, Manitoba, Canada.

Wilson, Edward O., 1998, *Consilience: The Unity of Knowledge*, Random House, New York.

Wilson, Fred L., 1968, "Fermi's theory of beta decay", *American Journal of Physics*, 36(12), pp. 1150–60.

Yoma (The Day), fifth tractate of the order *Moed* (Festivals) in the *Mishnah*.

Zadeh, L. A., 1965, "Fuzzy sets", *Information and Control*, vol. 8, pp. 338–53.

Index